HANDBOOK OF ELECTRONIC MATERIALS
Volume 3

HANDBOOK OF ELECTRONIC MATERIALS

Compiled by:
ELECTRONIC PROPERTIES INFORMATION CENTER
Hughes Aircraft Company
Culver City, California

Sponsored by:
AIR FORCE MATERIALS LABORATORY
Air Force Systems Command
Wright Patterson Air Force Base, Ohio

Volume 1:
OPTICAL MATERIALS PROPERTIES, 1971

Volume 2:
III-V SEMICONDUCTING COMPOUNDS, 1971

Volume 3:
SILICON NITRIDE FOR MICROELECTRONIC APPLICATIONS, PART I: PREPARATION AND PROPERTIES, 1971

In preparation:

Volume 4:
NIOBIUM ALLOYS AND COMPOUNDS

Volume 5:
GROUP IV SEMICONDUCTING COMPOUNDS

Volume 6:
SILICON NITRIDE FOR MICROELECTRONIC APPLICATIONS, PART II: APPLICATIONS

HANDBOOK OF ELECTRONIC MATERIALS
Volume 3

Silicon Nitride for Microelectronic Applications

Part 1
Preparation and Properties

John T. Milek

Electronic Properties Information Center
Hughes Aircraft Company, Culver City, California

IFI/PLENUM · NEW YORK-WASHINGTON-LONDON · 1971

This document has been approved for public release and sale; its distribution is unlimited. Sponsored by: Air Force Materials Laboratory, Wright-Patterson Air Force Base, Ohio.

Library of Congress Catalog Card Number 76-147312
ISBN 978-1-4684-6164-0 ISBN 978-1-4684-6162-6 (eBook)
DOI 10.1007/978-1-4684-6162-6

Softcover reprint of the hardcover 1st edition 1971
Plenum Publishing Corporation
227 West 17th Street, New York, N.Y. 10011

PREFACE

The large amount of literature on the technology of thin film silicon nitride indicates the interest of the Department of Defense, NASA and the semiconductor industry in the development and full utilization of the material.

This survey is concerned only with the thin film characteristics and properties of silicon nitride as currently utilized by the semiconductor or microelectronics industry. It also includes the various methods of preparation. Applications in microelectronic devices and circuits are to be provided in Part 2 of the survey. Some bulk silicon nitride property data is included for basic reference and comparison purposes.

The survey specifically excludes references and information not within the public domain.

ACKNOWLEDGEMENT

This survey was generated under U.S. Air Force Contract F33615-70-C-1348, with Mr. B.R. Emrich (MAAM) Air Force Materials Laboratory, Wright-Patterson Air Force Base, Ohio acting as Project Engineer. The author would like to acknowledge the assistance of Dr. Judd Q. Bartling, Litton Systems, Inc., Guidance and Control Systems Division, Woodland Hills, California and Dr. Thomas C. Hall, Hughes Aircraft Company, Culver City, California in reviewing the survey.

CONTENTS

INTRODUCTION

Literature Review

The Electronic Properties Information Center, with its world-wide literature coverage of all materials having a potential application in the field of electronic technology, was utilized for this survey. The availability of the Center's computer storage and retrieval system eliminated the need to conduct a formal literature search. Over 134 references were reviewed for this survey. The Center has previously prepared interim reports dealing with bulk silicon nitride (1) and thin film silicon nitride capacitors (2).

From the references accumulated for this survey, the chronology given in Table I-1 provides a brief synopsis of development in silicon nitride thin film technology and several review publications.

Table I-1. Developments in Silicon Nitride Thin Film Technology.

Date	Development	Reference
1960 (Feb.)	Thin film high temperature capacitor research and development	3
1962 (Dec.)	Vacuum deposited silicon nitride dielectric films	4
1965	Preparation and deposition of thin film silicon nitride for device application.	5
1966 (Jan.)	Use of Si_3N_4 for diffusion mask, passivation, gate insulator	6
1966 (Feb.)	Si_3N_4 for Microelectronic Applications Review	7
1966 (July)	Reported on properties of amorphous Si_3N_4	8
1966 (July)	Reported on diffusion masking with Si_3N_4	9
1966 (Oct.)	Symposium on silicon nitride	10
1967 (June)	Texas Univ. report on charge transport mechanisms, chemical vapor deposition process, electrical props. of Si_3N_4	11
1969	Thin film preparation by chemical vapor deposition review	12

Bulk Characteristics

Bulk silicon nitride has been commonly available for many years. It is manufactured by standard metallurgical techniques based on reacting silicon powder (13) with nitrogen at elevated temperatures (above 1300°C). P. Popper and S.N. Ruddlesden have reviewed the preparation, properties and structure of bulk silicon nitride.

Bulk silicon nitride has found widespread use as a refractory material in high temperature ceramic applications (14). It is a hard refractory with a use temperature of about 1200°C. It also is a good electrical insulator with reported resistivities of 10^{12} ohm-cm at room temperature and 10^{6} ohm-cm at 1000°C. Its thermal expansion coefficient is 2.5 x 10^{-6}/°C over the range of 20-1000°C. As a result of this low thermal expansion, its thermal shock resistance is very good. Wells (15) at the Lawrence Radiation Laboratory, University of California has surveyed this bulk material as a high-temperature radome material.

Technology Overview

A considerable volume of information has been generated on thin film silicon nitride in the past five years. In evaluating this literature, it became evident that the technology embraces a wide range of raw materials, processes, film structures, film properties, and numerous device applications. A complicating factor arose in evaluating the literature which involved distinguishing data on the making of devices to evaluate or elucidate the film characteristics from data where the end objectives are to develop specific devices or circuit components. Figure I-1 is an overview of all the various ramifications involving thin film silicon nitride and its combinations with silicon oxide. It points out the many variables inherent in the processing operations with the result that property data must be identified with the respective process and conditions of formation to have any merit for comparison purposes or to be meaningful.

Increasing interest in silicon nitride for microelectronic applications has resulted from the fact that silicon dioxide is not an entirely satisfactory dielectric material. The disadvantages of silica are well recognized and identified in the literature. Silica has strong affinity toward water and has high permeability to water vapor and other gases. Moreover, the migration of alkali ions in silica films has been found to be responsible for the instability of insulated gate field-effect transistors.

Figure I-1. SILICON NITRIDE TECHNOLOGY OVERVIEW

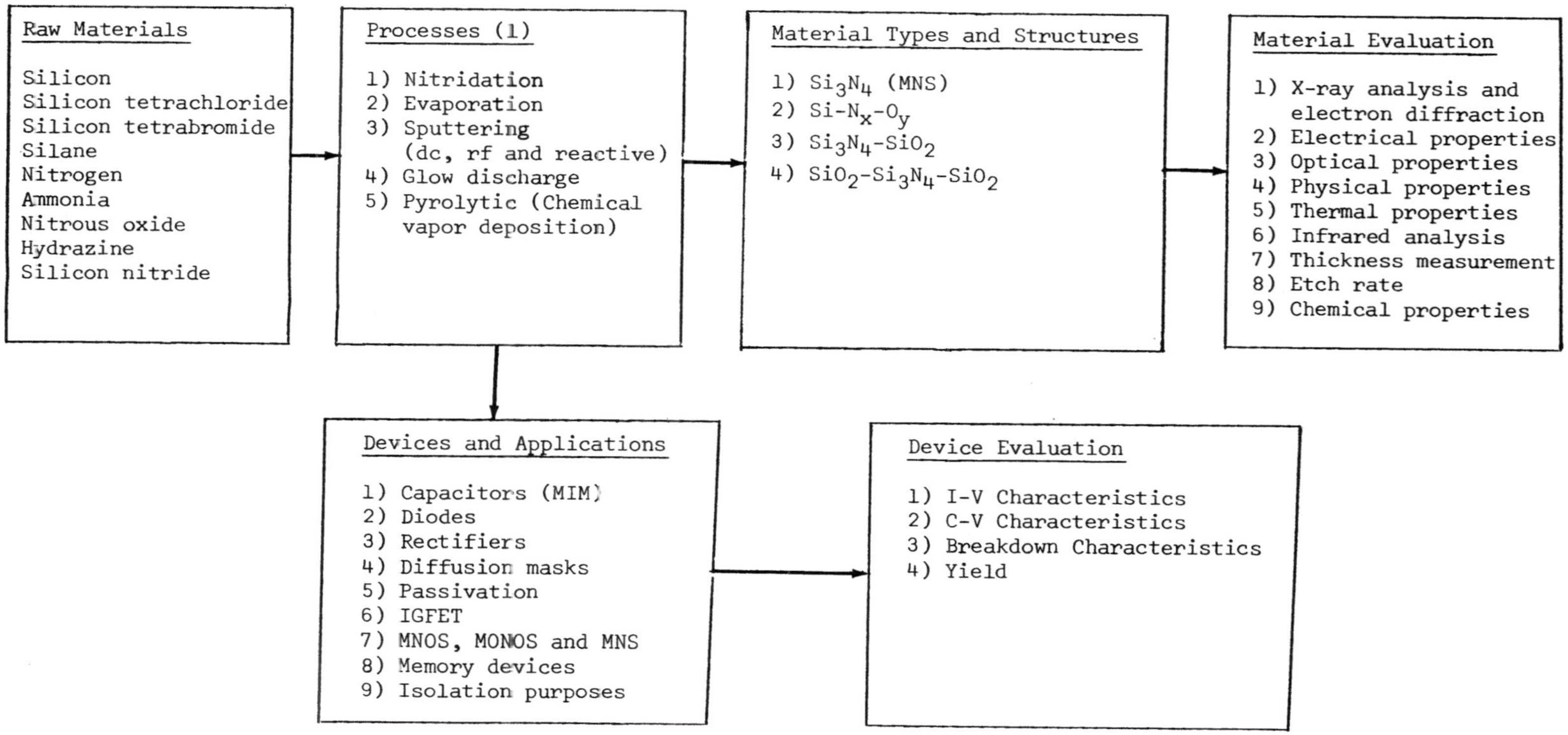

(1) Control Parameters: Reaction temperature, gas mixture ratios, pressure conditions, vacuum conditions, field voltages, rf power, substrate temperatures, flow conditions, growth rate of films, substrate preparation, and carrier gas used.

INTRODUCTION-REFERENCES

1. NEUBERGER, M.S. Collected Data and Bibliography on Silicon Nitride. HUGHES AIRCRAFT COMPANY, Electronic Properties Information Center, Culver City, Calif. IR-19, Feb. 1966. 41 p.

2. MILEK, J.T. Thin Film Dielectrics for Microelectronics, IR-63, July 3, 1968. 31 p.

3. BARNES, C.R. and C.R. GEESNER. Silicon Nitride Thin Film Dielectric. ELECTROCHEM. SOC., J., v. 107, no. 2, Feb. 1960. p. 98-100.

4. FELDMAN, C. and M. HACSKAYLO. Temperature Characteristics of Vacuum Deposited Dielectric Films. REV. OF SCIENTIFIC INSTRUMENTS, v. 33, Dec. 1962. p. 1459-1460.

5. STERLING, H.F. and R.C.G. SWANN. Chemical Vapour Deposition Promoted by r.f. Discharge. SOLID STATE ELECTRONICS, v. 8, no. 8, Aug. 1965. p. 653-654.

6. TOMBS, N.C. et al. A New Insulated-Gate Silicon Transistor. PROC. OF IEEE (Letters), v. 54, Jan. 1966. p. 87-89.

7. NEWMAN, R. et al. Silicon Nitride for Microelectronics. SPERRY ENGINEERING REVIEW, v. 19, no. 2, 1966. p. 2-9.

8. HU, S.M. Properties of Amorphous Silicon Nitride Films. ELECTROCHEM. SOC., J., v. 113, no. 7, July 1966. p. 693-698.

9. DOO, V.Y. Silicon Nitride, a New Diffusion Mask. IEEE TRANS. ON ELECTRON DEVICES, v. 13, 1966. p. 501-503.

10. ELECTROCHEMICAL SOCIETY, Symposium on Silicon Nitride, Philadelphia, Penn. Oct. 1966.

11. YEARGAN, J.R. and H.L. TAYLOR. Deposition of Silicon Nitride on Metal-Coated Substrates. 1967 SWIEECO Record, v. 19, pp. 16-4-1 to 16-4-8.

12. FEIST, W.M. et al. The Preparation of Films by Chemical Vapor Deposition. In: PHYSICS OF THIN FILMS, Edited by G. Hass and R.E. Thun, Academic Press, New York, 1969, pp. 237-322.

13. POPPER, P. and S.N. RUDDLESDEN. The Preparation, Properties and Structure of Silicon Nitride. BRITISH CERAMIC SOCIETY, TRANS., v. 60, Sept. 1961. p. 603-624.

14. PARR, N.L. Silicon Nitride-A New Ceramic for High Temperature Engineering and Other Applications. RESEARCH APPLIED IN INDUSTRY, v. 13, 1960. p. 261-269.

15. WELLS, W.M. Silicon Nitride as a High-Temperature Radome Material. University of California, Lawrence Radiation Laboratory, Livermore, Calif. Contract No. W-7405-eng-48, UCRL-7795. May 19, 1964. 19 p.

METHODS OF PREPARATION

Introduction

Silicon Nitride (Si_3N_4) films can be prepared by several different deposition techniques:

a) Direct nitridation

b) Evaporation

c) Glow discharge (dc and rf)

d) Sputtering (dc, rf and reactive)

e) Pyrolytic (chemical vapor deposition or CVD)

Of the methods shown in Figure II-1, the reactive sputtering and pyrolysis methods have been most frequently utilized or explored (based on the number of literature citations found in this survey). In each of these deposition methods, several parameters can be varied: temperature, flow rate, plasma density, pressure or degree of vacuum, ratio of reactants or electric field. Prior to deposition, the substrate silicon wafers are usually given a mechanical lap followed by a mechanical polish and in some cases a chemical polish. Wafers are also given typical semiconductor preoxidation cleaning immediately before silicon nitride film deposition. Heat treatment of the films is also utilized

Dalton and Drobek (1) and numerous other investigators have evaluated some of these above methods and varied the deposition technique conditions to obtain slightly different physical properties. They have also detailed the advantages and disadvantages of silicon nitride as compared to silicon dioxide.

Table II-1 lists reactions for forming the nitride that are thermodynamically feasible.

Table II-1. Free Energy of Over-all Reactions* (2)

Reaction	$\Delta G°$ kcal/mol 298°K	1300°K
$3Si + 2N_2 \rightarrow Si_3N_4$	−154.8	−71.4
$3Si + 4NH_3 \rightarrow Si_3N_4 + 6H_2$	−138.8	−165.4
$3Si + 2N_2H_4 \rightarrow Si_3N_4 + 4H_2$	−230.8	−257.8
$3SiCl_4 + 2N_2 + 6H_2 \rightarrow Si_3N_4 + 12HCl$	+15.3	−15.0
$3SiCl_4 + 4NH_3 \rightarrow Si_3N_4 + 12HCl$	+31.3	−109.0
$3SiCl_4 + 2N_2H_4 + 2H_2 \rightarrow Si_3N_4 + 12HCl$	−61.9	−202.6
$3SiH_4 + 2N_2 \rightarrow Si_3N_4 + 6H_2$	−194.4	−179.1
$3SiH_4 + 4NH_3 \rightarrow Si_3N_4 + 12H_2$	−170.4	−283.1
$3SiH_4 + 2N_2H_4 \rightarrow Si_3N_4 + 10H_2$	−270.4	−365.5

* JANAF Thermochemical Tables, Dow Chemical Company.

Arizumi and coworkers (3) at Nagoya University also made an analysis and comparison of the growth rates of silicon nitride employing thermodynamic calculations for the SiH_4-NH_3-H_2 and $SiCl_4$-NH_3-H_2 systems. All their theoretical predictions were experimentally supported and good agreement indicated that the main reaction in the former system was:

$$3SiH_4 + 12H_2 = Si_3N_4 + 4NH_3$$

In the $SiCl_4$-NH_3-H_2 system the agreement between theoretical results and experiments was rather poor which suggested a deviation from a quasi-equilibrium condition as well as the existence of complex chemical intermediates in the reaction.

The various deposition processes are discussed in detail in this section. Advantages and disadvantages of each process are enumerated. As an example, SiH_4 (silane) has a lower deposition temperature (710°C) than that for $SiCl_4$ (silicon tetrachloride) (850-890°C) for general production Si_3N_4 operations. Additionally, Kohler (4) has presented interesting comparative data on the $SiCl_4/NH_3$ process vs the SiH_4/NH_3 process.

Property data pertaining to each process is detailed in the properties section of this report.

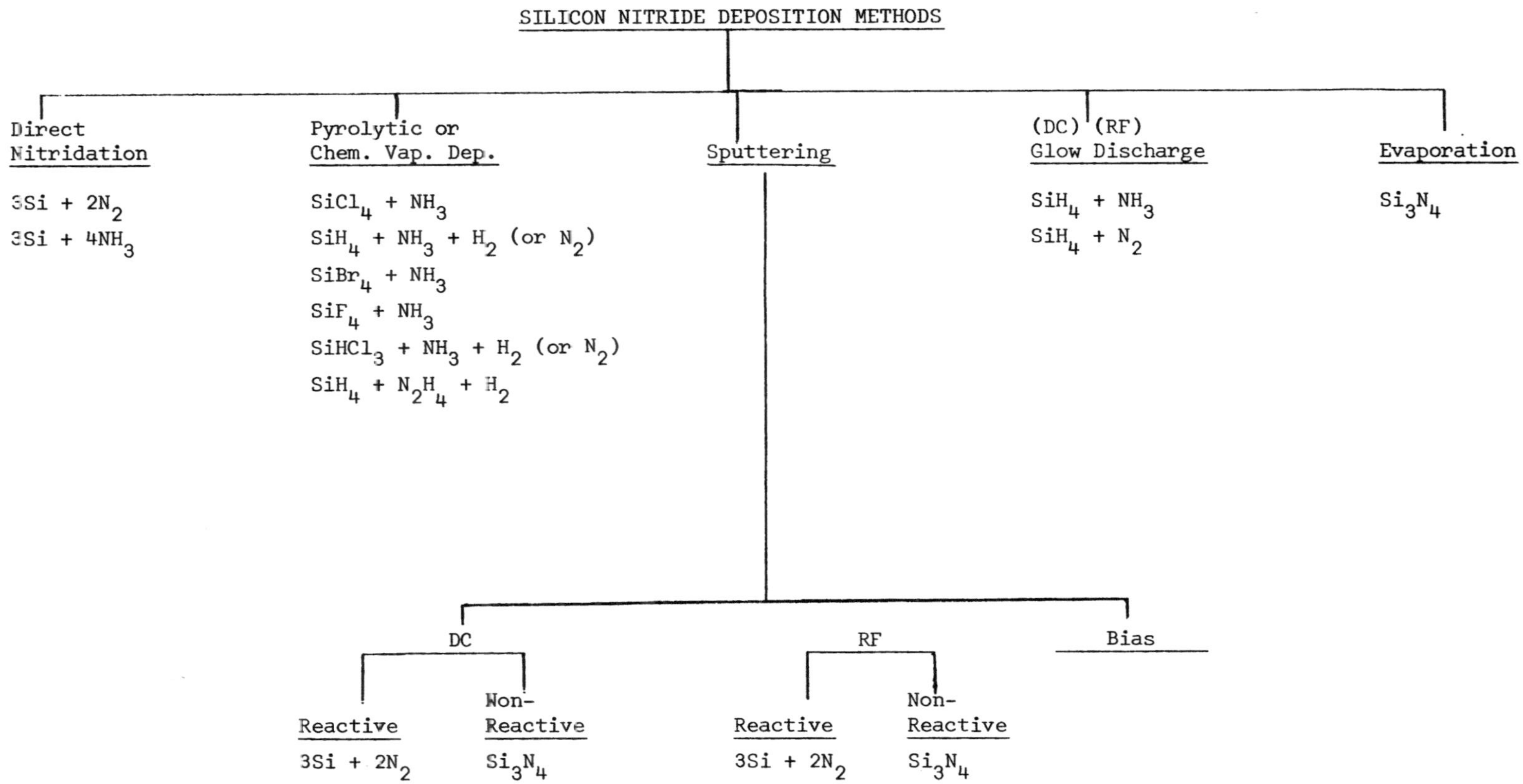

Figure II-1. Silicon Nitride Deposition Techniques.

Direct Nitridation Method

Early attempts at Sprague Electric Co. (5) led to the conclusion that direct nitridation is possible, but very hard to control - probably because the experiments were conducted at 1250°C, employing low nitrogen flow rates (1 liter/mm). Film formation rates were very slow (approx. 500 Å/16 hours) and their refractive index varied from 2.0 to 2.4. Later studies indicated that the growth by direct nitridation was self-limiting. Even glow discharge experiments with atomic nitrogen showed it was virtually impossible to diffuse nitrogen into silicon to any extent at 1200°C, but in all cases formed a thin silicon nitride layer, which effectively stopped any nitrogen diffusion.

Kaiser and Thurmond (6) had earlier carried out experiments with thermal nitridation at 1300°C and plasma anodization and came to the same conclusion. The nitridation of silicon is a slow process at 1300°C and above because of the diffusion-controlled mechanism. Furthermore, the heating of single crystal silicon wafer of {111} orientation in nitrogen or an argon-nitrogen mixture has been found to produce discontinuous films of crystalline alpha-silicon nitride (7). Such films are not suitable for many device purposes because of grain boundaries.

The direct nitridation reactions are expressed as follows:

$$3Si + 2N_2 = Si_3N_4$$

$$3Si + 4NH_3 = Si_3N_4 + 6H_2$$

Turkadogan and coworkers, as early as 1958, reported on their preparation of silicon nitride thermally by direct combination of the elements and reported on the physicochemical properties of the compound. (8)

The direct nitridation of a Si (111) surface by the LEED (low energy electron diffraction) technique has also been studied. Using N_2 and NH_3 as the nitriding medium, the LEED method was used to provide information about the structure of the substrate surface and of the nitride layers at the interface, as they formed. It was found that between 700°C and 1200°C, the initial nitridation resulted in the growth of ordered layers which showed a definite epitaxial relationship with the silicon substrate. Continued nitridation led to more irregular growth of nitride, which has so far prevented electrical assessment of the interface.

Single crystal silicon heated in nitrogen at 1300-1350°C yielded alpha-silicon nitride in the form of dark brown needles, ribbons, or platelets. (10) The film obtained by this method, could not be etched with 50% HF or 1 part 50% HF + 3 parts HNO_3 at room temperature.

Films of silicon nitride were also grown in a nitrogen atmosphere by Kendall (11) at a temperature of about 1000°C to 1200°C. The growth rate was found to be quite rapid; however the film so formed consisted of microcrystallites and was uneven and hence unsuitable for commercial use. He found the concentration of surface states N' to be approx. $10^{11}/cm^2$. The film (1500 Å thick) was found to be quite stable in atmospheres of wet and dry oxygen and nitrogen at temperatures of up to 100°C.

Evaporation Method

In 1962, Feldman & Hacskaylo (12) prepared 1-micron thick silicon nitride dielectric films by vacuum evaporation of the nitride at 2100°C onto a fused silica substrate (temperature = 250°C). The measured dielectric constant differed considerably from previously reported pyrolytic deposited material.

Electron beams have been used to evaporate silicon nitride film at pressures of 10^{-5} to 10^{-6} torr. With the substrate temperature maintained above 200°C, the resulting dielectric films were found to be stable and transparent. X-ray analysis showed the films (200-2000 Å thick) to be amorphous in structure. (13)

Kendall (11) attempted the straightforward vacuum evaporation of silicon nitride using a pressure of about 10^{-9} torr and a fused quartz crucible with a platinum heater. The nitride was deposited on to the (111) face of 7 ohm-cm n- and p-type silicon and an aluminum field plate or gate was deposited on the nitride. The resultant films were inhomogeneous, containing Al filaments which were presumably formed because of pinholes in the original nitride layers. He was able to determine N' (concentration of surface states) which was found to be about 10^{10} cm^{-2}.

Glow Discharge Method

The physical processes involved in the glow discharge deposition process are not simple since they involve the interaction of positive ions, electrons, and excited molecules as well as the presence of neutral atoms and molecules. Moreover, the glow or "cold" discharge occurs between the Townsend and the arc discharge which are only part of the family of electrical breakdown phenomena. Discharges can be further classified into two different types: (a) capacitive or "E" type (e.g. excited by an electric field and (b) inductive of "H" type (e.g. excited by a magnetic field).

Silicon nitride has been prepared by the glow discharge process involving the mixing and reaction of silane (SiH_4) and ammonia (NH_3); other reactants can be used as well. Both dc and rf glow discharge methods have been explored. In this technique, the high temperature generated within the gaseous discharge, provides the energy required to decompose the reactant gases and facilitate film deposition while not requiring high ambient temperatures to which the substrate must be exposed. Generally, the rf glow discharge is preferred because it is a relatively simple system not requiring electrodes, targets, water cooling and biasing, thus eliminating any problems with electrode contamination through the use of high purity gaseous sources. The energy for the reaction is supplied by a rf coil surrounding a fused silica reaction tube through both inductive and capacitive coupling to a low pressure gas stream (typically 10^{-1} Torr) as shown in Figure II-2.

In 1965, Sterling and Swann (14) reported a method for depositing films of silicon nitride by the rf discharge process. Savage et al. (15) at the Radar Research Establishment in England also investigated the glow discharge technique and noted that the properties of the films depended on a number of parameters including substrate temperature, rf field conditions, ratio by volume of the reacting gases and reaction chamber geometry. In their experiments, they found that the silicon nitride deposited films developed random circular mounds which cracked around the edges, leaving a circular hole. Poor adhesion of the deposited film, plus strain caused by the difference in temperature coefficients of the silicon nitride and the silicon, were the probable cause of this effect.

It was found that the physical and electrical properties (including dielectric breakdown strength) of silicon nitride films deposited by reacting silane and ammonia in a rf glow discharge could be correlated with changes in concentration of gas mixture components (16). The apparatus used to deposit these films at a fixed substrate temperature at 300°C is shown in Figure II-2.

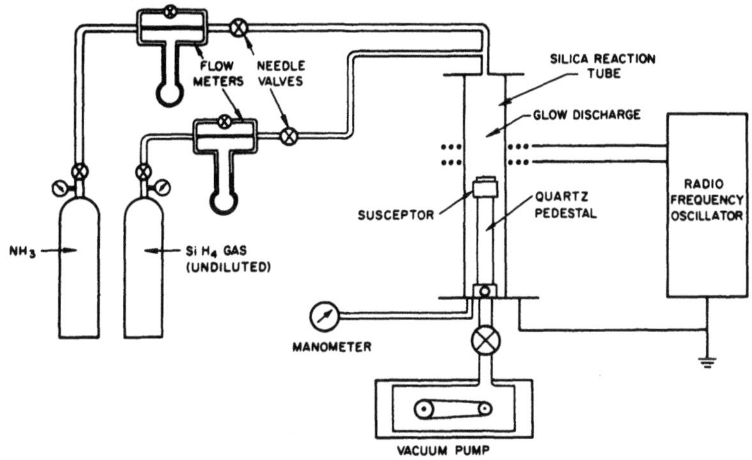

Figure II-2. Glow discharge apparatus. (16)

The reason for the 300°C substrate temperature is because of the superior etch resistance of the films formed at this temperature. Using ammonia and silane in a rf glow discharge deposition, growth rates of 35 to 180 Å/min. have been found to vary directly with silane concentration from 10 to 15% silane. The dielectric constant property varied directly with silane concentration while etch rate varied inversely with it.

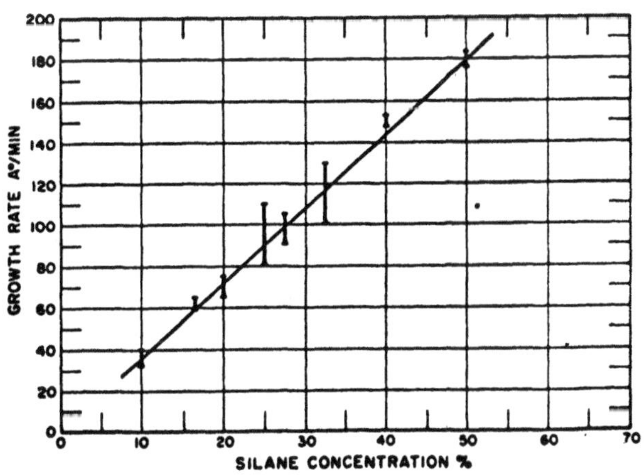

Figure II-3. Growth rate vs. silane concentration. (16)

Figure II-3 shows the rate of film deposition vs silane concentration under conditions of constant radio frequency power, gas flow rate, and substrate temperature. The growth rate is substantially linear, having a rate of 180 Å/min at 50% silane concentration and 35 Å/min at 10% concentration. One of the problems in this deposition method is the purity of ammonia. Difficulty has been experienced in removing water from the ammonia.

Silicon nitride can also be produced by a rf (27 MHz) glow discharge reaction of silane (SiH_4) and nitrogen (N_2) as a substitute for ammonia. (17) This eliminates the problem of water contained in ammonia. One part of silane was mixed with 10^3 parts of nitrogen. The mixed gas pressure of SiH_4 and N_2 was changed in the range from 0.15 mm Hg to 1.1 mm Hg. Silicon nitride was deposited on a chemically etched n-type silicon wafer (resistivity of 10 ohm-cm). Growth rates of silicon nitride deposited by the glow discharge reaction at the mixed gas pressure of 0.85 mm Hg are on the order of 40 Å/min. Etch rates of this nitride in 24% HF at room temperature are on the order of 50 Å/min. Etch rates of Si_3N_4 deposited by the glow discharge reaction of SiH_4 and N_2 are on the same order with that of Si_3N_4 deposited by the glow discharge reaction of SiH_4 and NH_3.

Gereth, Holzer and Scherber (18) in Germany also deposited Si_3N_4 films in an rf glow discharge at temperatures below 400°C using SiH_4 and N_2. They found the properties depended strongly on the growth conditions. Basically, the structure of the nitride layer varied with the SiH_4 concentration. At a SiH_4 concentration of 0.1 and 13%, respectively, the refractive index is 1.95 and 2.9; the breakdown voltage 10^7 and 5×10^6 V/cm, and the etch rate in 20% HF 310 and < 10 Å/min. The electrical properties are studied by capacitance measurements on MNS diodes. The charge density derived from the flat-band voltage is positive and lies typically at $10^{12}/cm^2$ on n-type Si substrates. The value is twice as high on p-type Si. No relationship between flat-band charge and SiH_4 concentration was observed. Growth temperatures below 250°C drastically increase the flat-band charge. The C-V curve of these films on Si shows the following hysteresis effect: the flat-band voltage follows the applied bias. The magnitude of the hysteresis depends on the maximum applied field strength, but the hysteresis disappears below a certain threshold of about 1 to 2×10^6 V/cm. On the other hand, the magnitude of the hysteresis strongly decreases with increasing SiH_4 concentration. Films prepared below 300 and 400°C show minimum hysteresis effect.

Ion Beam Method

Aisenberg and Rohatgi (19) reported on their inability to deposit silicon nitride films by using an ion source in conjunction with a silicon electrode and nitrogen gas plasma. It was hoped that a beam of silicon and nitrogen ions would be extracted from the source and made to impinge on the deposition substrate. Lack of time and funds prevented further exploration of this deposition method which had been shown to work with single silicon ion components.

Sputtering Methods

Dc, rf and reactive sputtering techniques have been widely explored for preparing silicon nitride films. In rf reactive sputtering, the cathode is a high-purity block of silicon or silicon nitride. High energy electrons passing through a vacuum system (back filled with a small amount of ammonia and/or nitrogen in argon) generate heavy argon ions. The heavy ions are then attracted to the silicon or silicon nitride cathode. Secondary emission results due to the collision of the heavy ions with the surface of the cathode. Silicon or silicon nitride ions removed from the surface are then attracted to the anode which is the silicon wafer that is to be coated with the barrier layer. Comparative studies of the various sputtering methods have shown that the rf reactive sputtering is superior because of its freedom from dissociation problems of direct sputtering, as well as the film damage caused by electron and/or negative ion bombardment resulting from higher dc anode-cathode voltages in dc reactive sputtering. Moreover, oxygen contamination has been found to be more of a problem with reactive than with direct sputtering since it is more reactive than the nitrogen normally used as the reactive gas.

The advantages of this method include the fact that the substrate temperature can be low compared to the substrate temperature involved in the chemical vapor deposition technique. This allows the films to be deposited on substrates which would melt at the temperature required for chemical vapor deposition, (e.g. Al). Another advantage of the low substrate temperature is that the difference in coefficient of expansion is less important because the expansion is much less than for chemical vapor deposition. The major disadvantage of rf sputtering is that the concentration of the two constituents cannot be varied with ease and the stoichiometry of the films is less controllable.

Janus and Shirn (20) at Sprague Electric Co. have actively investigated reactively-sputtered silicon nitride films using both the glow-discharge and supported-glow-discharge reactive sputtering techniques. They found that adherent amorphous silicon nitride films can be deposited on a variety of substrate materials at relatively low temperatures using the reactive sputtering technique. The supported glow discharge approach was found to be superior to glow-discharge sputtering for the reproducible preparation of uniform, stoichiometric films. These investigators prepared transparent adherent films in the thickness range of 300 to 15,000 Å. The relationship between sputtering voltage and deposition in the current range of 90 to 120 mA is shown in Figure II-4.

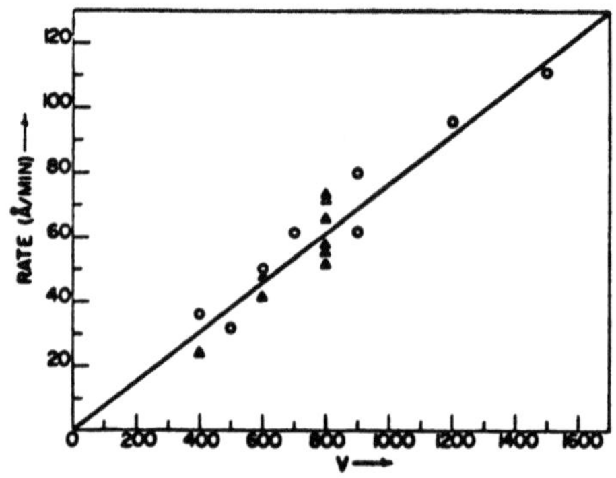

Figure II-4. Deposition rate of silicon nitride as a function of sputtering voltage. (20) △ 7059 glass substrate
O Si substrate

According to Cordes (21), radio-frequency reactive sputtering can provide nitride films with higher deposition rate and less oxide contamination as compared with dc reactive sputtering. Furthermore, films obtained by dc reactive sputtering have been found to contain excess silicon on the basis of dissolution rate, refractive index, uv transmission, and dc conductivity measurements. Silicon nitride films deposited by rf reactive sputtering on silicon substrates at relatively low temperatures, 250°C, have been reported to be suitable for most device purposes.

In his study of sputtered silicon nitride films, Hu (22) found that frequently one also obtains sputtered films with surface smoothness comparable to pyrolytic films. Typical films prepared by means of d-c reactive sputtering are usually rougher. Since films with this type of surface texture were found to be leaky in moderate electric fields, he assumed that submicroscopic cracks exist in these d-c sputtered films. The grain texture became finer and less pronounced as the d-c voltage decreases.

Pyrolytic Methods

The pyrolytic deposition process has been the most widely investigated and used method of preparing silicon nitride films. Chemical reagents used in the process are introduced as vapors into a chamber maintained at an elevated temperature. These vapors are cracked or decomposed to deposit a nitride layer or film onto a substrate located within the chamber prior to the introduction of the gaseous reactants. Various reactants or combinations have been explored to date:

a) Silane (SiH_4) and ammonia (NH_3):
$$3SiH_4 + 4NH_3 \longrightarrow Si_3N_4 + 12H_2$$

b) Silicon tetrachloride ($SiCl_4$) and ammonia (NH_3):
$$3SiCl_4 + 4NH_3 \longrightarrow Si_3N_4 + 12HCl$$

c) Silicon tetrafluoride (SiF_4) and ammonia (NH_3):
$$3SiF_4 + 2NH_3 \longrightarrow Si_3N_4 + 2NH_4F + 10\ HF$$

d) Silane (SiH_4) and hydrazine (N_2H_4):
$$SiH_4 + N_2H_4 \longrightarrow Si_3N_4 + NH_3 + H_2 \text{ (not balanced)}$$

e) Trichlorosilane ($SiHCl_3$) and ammonia (NH_3)

The use of the pyrolytic process for semiconductors is advantageous since it is compatible with the epitaxial reactors already in use by most manufacturers (see Equipment Section).

The main process variables affecting the deposition rate and the resulting dielectric film properties as well as uniformity and reproducibility are deposition temperature, concentration and purity of gaseous reactants, the reactants employed, the deposition chamber geometry, and the total gas flow rate. Additionally, film properties such as refractive index (n), etch rate, density, chemical composition, and surface state density (Q_{ss}) can be made to vary widely.

Silane and Ammonia Reaction

Hu (22) and Doo et al (23,24) reported on the pyrolytic deposition of amorphous silicon nitride films by the reaction of silane (which starts to decompose at 500°C) and ammonia and determined the property changes as a function of reactant composition (SiH_4:NH_3).

The potential of developing cracks in films deposited on silicon was greatly reduced by decreasing the ammonia injection ratio. However, all films with low NH_3 and previously high NH_3 types were found to be amorphous. Decreasing the NH_3 injection rate increases the index of refraction and the electronic leakage while decreasing the dissolution rate (in 48% HF). They also found the deposition rate, the dielectric constant and the flat band charge density are not significantly affected by the NH_3 injection rate. There was quite a difference in the deposition rate when hydrogen versus nitrogen was utilized as the carrier gas as shown in Figures II-5 and II-6.

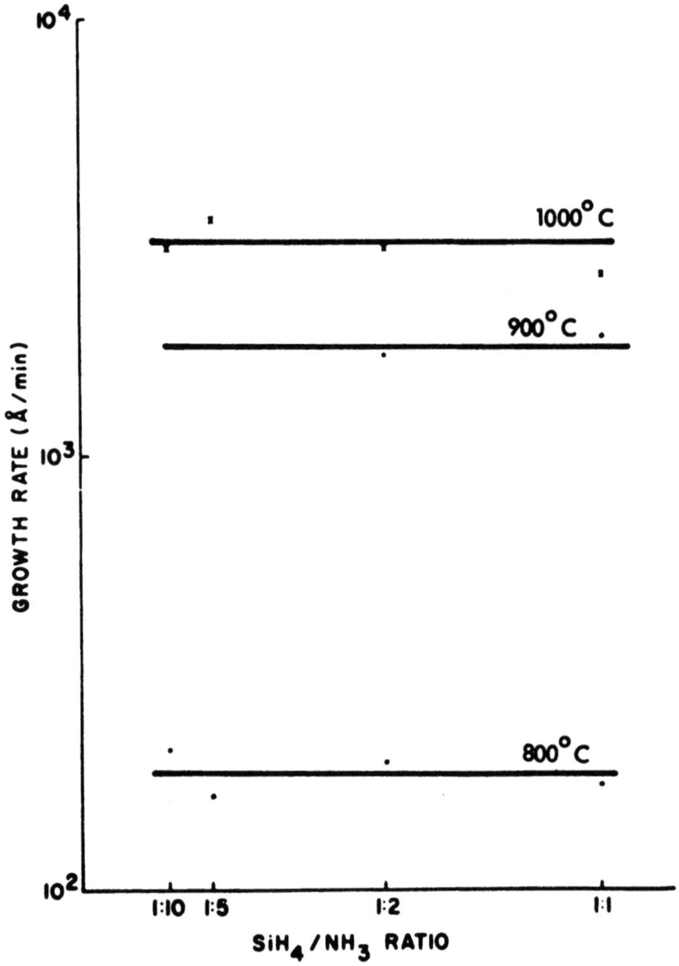

Figure II-5. Growth rate vs. $SiH_4:NH_3$ ratio carrier gas H_2. (25)

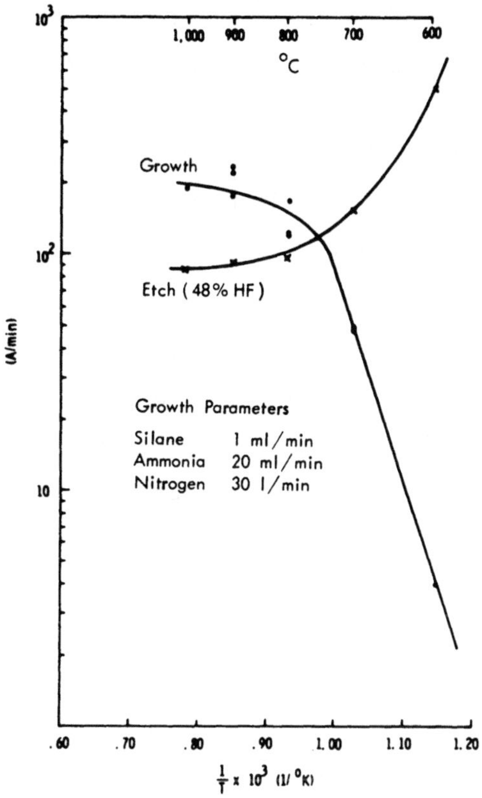

Figure II-6. Growth rate vs. temperature and etch rate vs. temperature. (25)

In the hydrogen carrier gas approach, the growth rate of the silicon nitride film at a given temperature is practically independent of the ammonia injection rate with the SiH_4:NH_3 ratios of 1:1 to 1:10. For the nitrogen carrier gas experiments, the growth rate at temperatures above 800°C increases slightly with temperature up to 1000°C while between 650-750°C, it increases rapidly with temperature. Limited investigations have been made using helium and argon carrier gases for silicon nitride film deposition. The electrical properties of the film grown in helium are superior to that grown in any other carrier. The argon films were inferior to those grown in nitrogen or helium.

The substrates were n-type silicon wafers of about 1.5-2.5 ohm-cm, chemically polished in HF-HNO_3 solution. Substrate temperatures varied from 800 to 1000°C. The maximum growth rate occured at 1000°C; cracking and warping problems were also encountered. Films deposited below 900°C appeared amorphous and showed no x-ray diffraction lines. Between 900° and 1000°C, some small crystallites grew over the surface and above 1100°C, they were almost continuous. In another paper, Doo & Kerr (25) used the same SiH_4/NH_3 system but added N_2 as a carrier gas to achieve better uniform growth. Besides, the SiH_4-NH_3-N_2 is much less hazardous to operate than the SiH_4-NH_3-H_2 system. They found that the electrical properties of the silicon nitride films grown in nitrogen are comparable to if not better than that in hydrogen.

Chu, Lee and Gruber (7) have also deposited amorphous silicon nitride by the nitridation of silane with ammonia on silicon substrates at 800 to 1100°C (metal-nitride-semiconductor structures were fabricated with this process). The most important control parameters determining the deposition rate of the silicon nitride films were found to be the substrate temperature as shown in Figure II-7 and the composition and flow rate of the reactant mixture shown in Figure II-8.

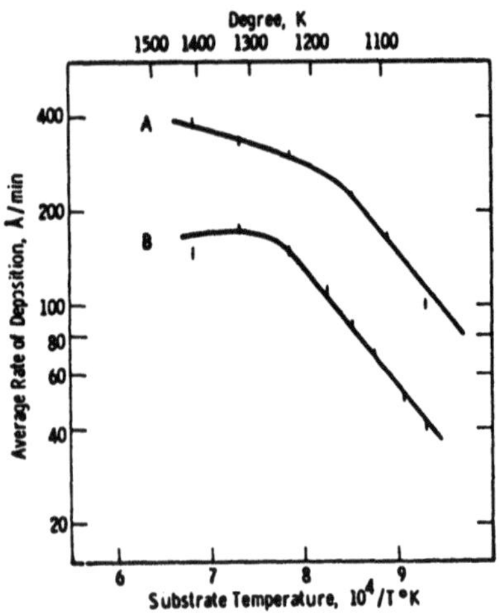

Figure II-7. Temperature dependence of deposition of silicon nitride films. A: Flow rate of NH_3 = 20 l/min, flow rate of SiH_4 = 2 ml/min. B: Flow rate of NH_3 = 40 l/min, flow rate of SiH_4 = 2 ml/min. (7)

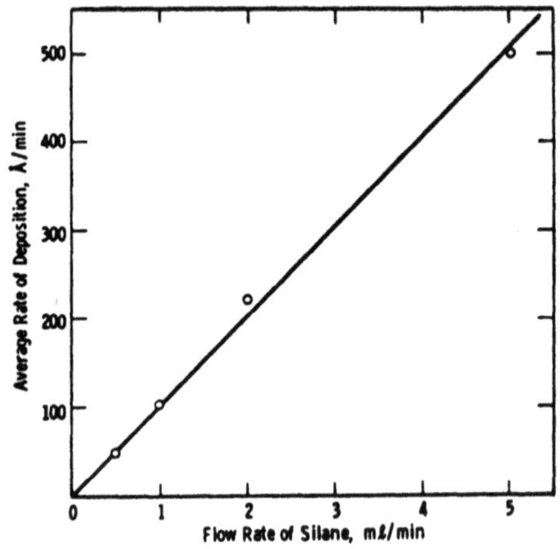

Figure II-8. Deposition rate of silicon nitride as a function of the silane flow rate. Flow rate of NH_3 = 20 l/min, substrate temperature = 900°C. (7)

Under these conditions, the deposition rate was 50 to 200 Å/min. The nitride films obtained in all cases were uniform, transparent, and adherent to the substrate, and were found to be amorphous by electron diffraction examinations. The crystallographic orientation of the substrate surface ({111} or {100}), the reactant composition were found to have no significant effects on the C-V relations.

Silicon nitride films have been deposited by the gas phase reaction between silane (SiH_4) and ammonia (NH_3) and also using nitrogen (N_2) as the carrier gas in a horizontal resistance-heated furnace at the temperature range of 650° to 950°C. (26) Infrared absorption data and data showing the dependence of the etching rate on deposition temperature were developed. The substrate was a phosphorus-doped n-type silicon wafer of approx. 5 ohm-cm resistivity with (111) surface and its surface was etched away about 30 microns by a HNO_3/HF etchant.

Yeargan & Taylor (27) at the University of Texas prepared silicon nitride films from the reaction of silane and ammonia in quartz reaction chamber using a resistive-heated graphite substrate heater. The thin films deposited on substrates of quartz and silicon exhibited a reproducible dc current component that occurs at high fields before breakdown. A strong temperature dependence of ln I on $V^{1/2}$, and an absence of electrode control indicated that the dominant mechanism at room temperature is thermal emission from the bulk of the nitride.

The film character of pyrolytic Si_3N_4 prepared by the SiH_4/NH_3 reaction as a function of temperature is shown in Figure II-9.

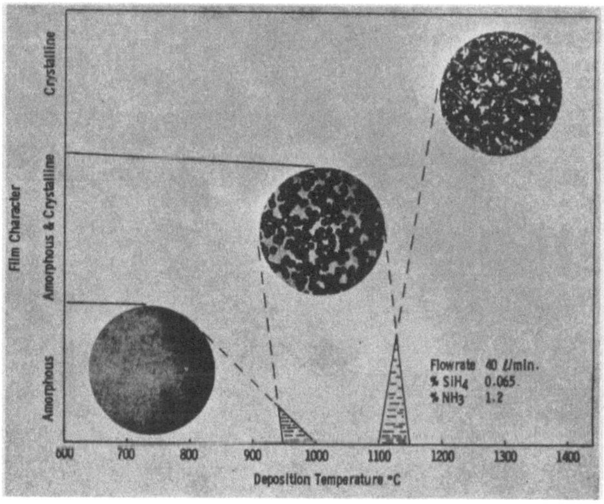

Figure II-9. Film character vs. deposition temperature. (28)

Figures II-10, II-11 and II-12 provide data on the effect of temperature on the deposition rate for various silane concentrations; as a function of ammonia concentration.

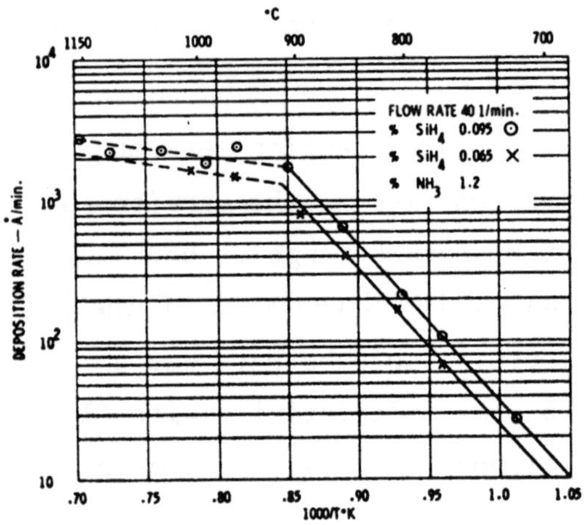

Figure II-10. Deposition rate vs. temperature. (28)

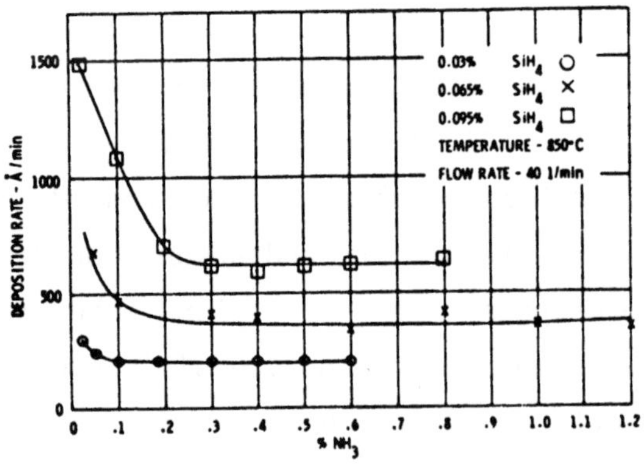

Figure II-11. Deposition rate vs. percent NH_3. (28)

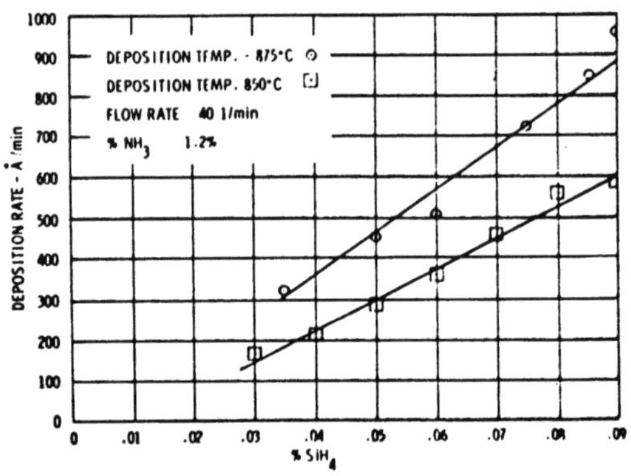

Figure II-12. Deposition rate vs. percent SiH_4. (28)

Amorphous silicon nitride layers (up to 5000 Å) were produced by Schaffer and Swaroop (29) by chemical vapor phase reaction between ammonia (NH_3) and silane (SiH_4) ultrapure gases in a hydrogen environment between 825° and 1150°C. Under the growth conditions used, the growth rate increased approximately linearly from nil at 825°C to 575 Å/min at 1150°C as shown in Figure II-13.

Figure II-13. Effect of growth temperature upon growth rate and etch rate of chemically vapor-deposited silicon nitride. (29)

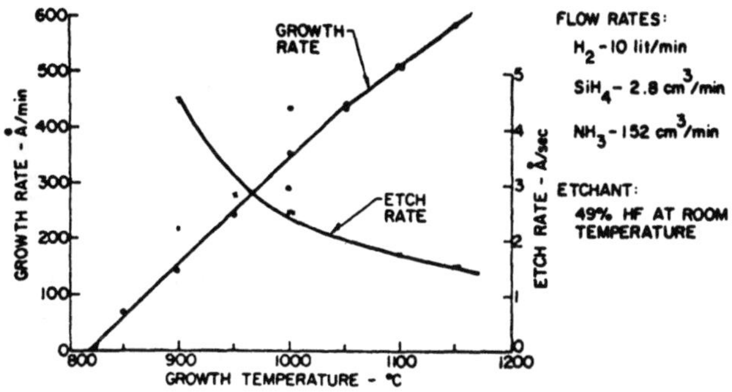

The etch rate in 49% HF decreased because of the higher density material obtained with increasing deposition temperature. These investigators reported that continuous amorphous films, free from macroporosity and cracks, were obtained up to layer thicknesses of approx. 5000 Å. Above this thickness, the wafers bowed noticeably and cracks were formed.

Kohler (4) has also investigated the variables involved in depositing silicon nitride by the SiH_4/NH_3 method. Figures II-14 and II-15 give deposition rate data as a function of each flow rate.

Figure II-14. Deposition rate of silicon nitride as a function of the NH_3 gas flow in the SiH_4 system. (4)

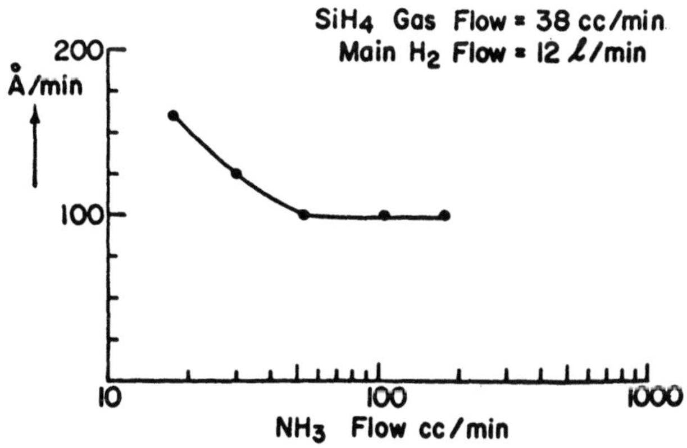

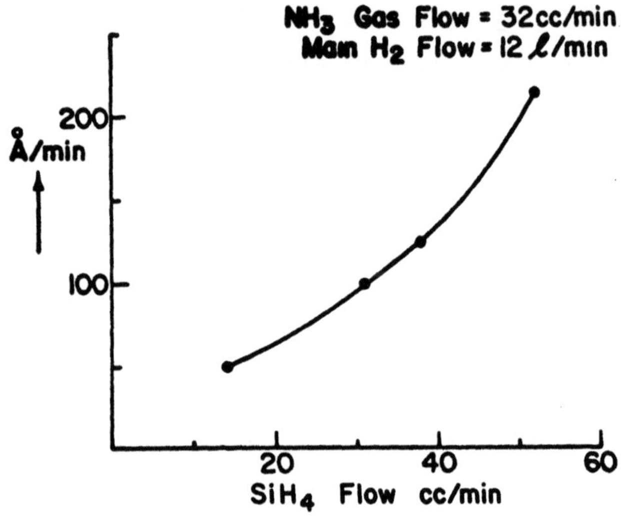

Figure II-15. Deposition rate of silicon nitride as a function of the SiH_4 gas flow rate. (4)

From these data, it is apparent that the growth rate in the SiH_4/NH_3 process is determined mainly by the SiH_4 gas flow.

Silicon nitride films were prepared by Carlson et al (30) using silane/ammonia in various ratios at 850°C and the resulting deposition rates, etching rates (in HF and H_3PO_4), refractive indices, dc resistivity and dielectric constant were evaluated. They concluded that films with the most desirable characteristics are formed at silane/ammonia ratios below 0.1. The constancy of all the characteristics in this low-silane region is a great asset in the establishment of a stable, reproducible process. Figure II-16 substantiates the conclusion.

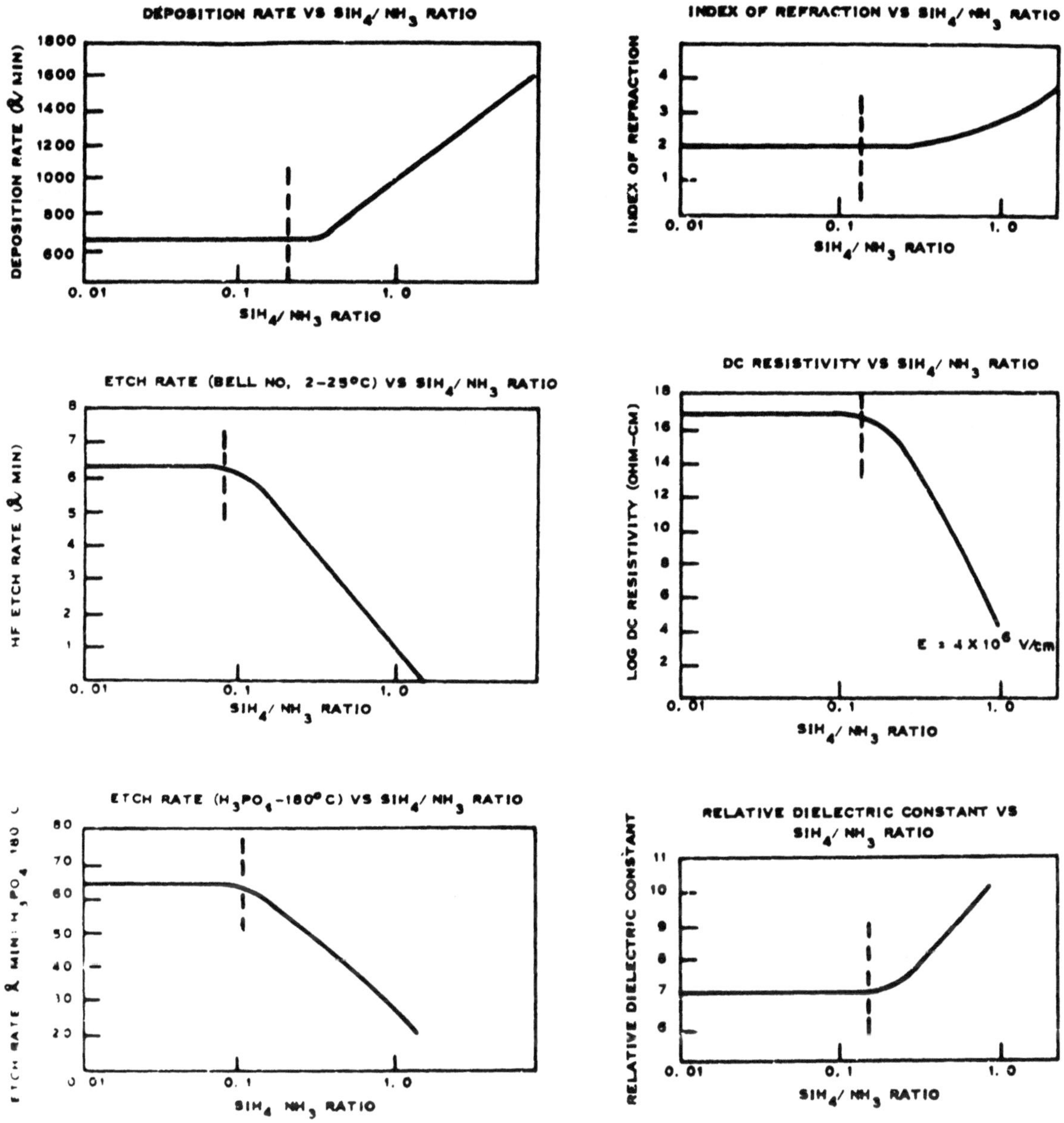

Figure II-16. Properties of silicon nitride. (30)

Kendall (11) coated 10 ohm-cm Si (n-type) with Si_3N_4 films by reacting ammonia and silane at elevated temperatures and found the proportion of SiH_4 to NH_3 to affect both the growth rate and the relative permittivity of the film. The temperature of deposition affected the crystallinity of the nitride and hence its thermal coefficient of expansion. Results with five parts by volume of SiH_4 to one of NH_3 and a substrate temperature of 850°C appeared to give the lowest density of interface states. The rf glow-discharge pyrolytic method produced similar films more consistently with a 3:1 ratio by volume and a substrate temperature of 350°C. Results from types of film were identical. The nitride deposited on the Si was 1150 Å thick, homogeneous, and had a relative permittivity of 8.7. Its resistivity was about 5 x 10^{14} ohm-cm and its dielectric strength at least 10^7 V/cm at room temperature.

Silicon Tetrachloride and Tetrafluoride Reaction

The ammonolysis of silicon tetrachloride through a complicated series of reactions can produce alpha-silicon nitride according to Chu et al (7). The nitride films were deposited on silicon substrates at 1000 to 1200°C using ammonia (NH_3) at flow rates of 6 to 16 lit/min. and silicon tetrachloride ($SiCl_4$) at flow rates of 3 x 10^{-4} to 3 x 10^{-3} moles/min. Under these conditions, adherent and transparent films were obtained and the deposition rate was found to be essentially independent of temperature in the temperature range studied as shown in Table II-2.

Table II-2. Deposition Rate of Silicon Nitride as a Function of Reactant Composition and Flow Rate. (7)

Flow rate of NH_3, l/min	$SiCl_4/NH_3$ molar ratio	Average deposition rate, μ/min
16	6 x 10^{-3}	0.8
12	6 x 10^{-3}	0.95
6	6 x 10^{-3}	1
6	3 x 10^{-3}	0.65
6	1.2 x 10^{-3}	0.5

Grieco et al (2) at Bell Telephone Labs. also studied the silicon tetrachloride ($SiCl_4$)/ammonia (NH_3) reaction in the presence of silicon substrates (heated in the temperature range 550°-1225°C) and reported on the kinetics of the deposition process. The films (amorphous) obtained ranged in thickness from 500 Å to 2 microns. Deposition rate as a function of silicon tetrachloride pressure and temperature is shown in Figures II-17 and II-18. His kinetic studies reinforce Chu's statement on the complexity of the $SiCl_4/NH_3$ reaction processes.

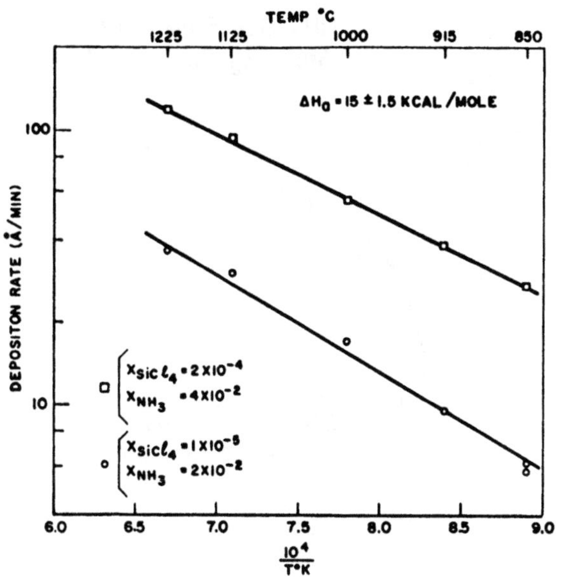

Figure II-17. Temperature dependence of silicon nitride deposition at 2×10^{-4}: 4×10^{-2} and 1×10^{-5}: 2×10^{-2} mol fractions, respectively, of $SiCl_4$:NH_3. (2)

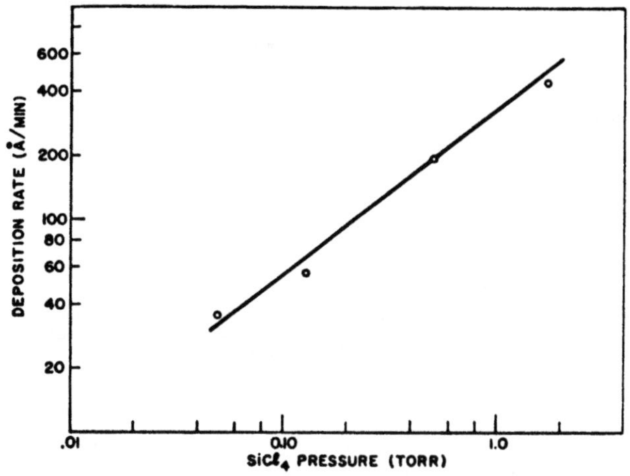

Figure II-18. Effect of $SiCl_4$ partial pressure on deposition rate (at 1000°C and NH_3 partial pressure of 28 Torr). (2)

Silicon nitride depositions were carried out in a horizontal rf quartz reactor at about 950°C by Deal et al (31). Hydrogen and ammonia were passed into the reactor along with $SiCl_4$ at relative flow rates: H_2 (4 lit/min.): NH_3 (4 lit/min.): $SiCl_4$ liquid (0.2 mil/min) or (approx. 40 ml/min. vapor) with a resulting deposition rate of approx. 0.12 microns/min.

Kohler (4) has investigated the variables involved in depositing silicon nitride by the $SiCl_4$/NH_3 reaction. Figures II-19 and II-20 gives the deposition rate data as a function of each gas flow rate.

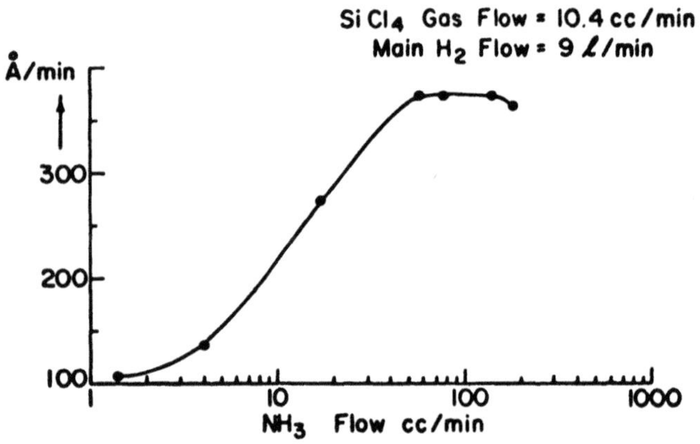

Figure II-19. Deposition rate of silicon nitride as a function of the NH_3 gas flow rate in the $SiCl_4$ system. (4)

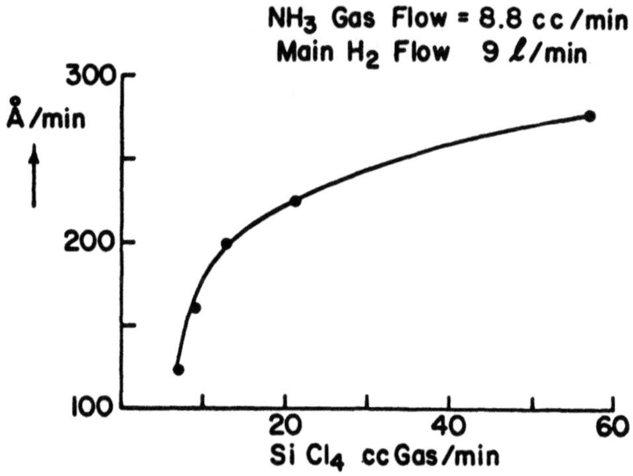

Figure II-20. Deposition rate of silicon nitride as a function of the $SiCl_4$ gas flow rate. (4)

From Kohler's data it is apparent that high ammonia flow rates result in high growth rates with the $SiCl_4/NH_3$ system, thus giving a less controllable process for the very thin films generally used for devices. Increasing the $SiCl_4$ gas flow rates has less effect on growth rates above some minimum value.

Chu (2) has reviewed the pyrolytic method and compared it with others:

1) The $SiCl_4/NH_3$ reaction involves several polymeric intermediates and is more difficult to control than the nitridation reaction.

2) The important parameters of the SiH_4/NH_3 deposition process are the substrate temperature and the composition of the reactant mixture. At substrate temperatures below 1000°C, the deposition rate is an exponential function of the reciprocal temperature. Silicon nitride films can be deposited by using only NH_3 and SiH_4 in the reactant mixture or by adding H_2 as a carrier gas. In the former case, NH_3 is present in large excess to insure that the deposited film is not nitrogen deficient. By adjusting the composition and flow rate of the reactant mixture, deposition rate can be varied from 10 to 1000 Å/min in the temperature range 700°-1000°C. When H_2 is used as the carrier gas for the reactant mixture, the NH_3/SiH_4 molar ratio in the reactant becomes important.

Kuntz (32) reported the preparation of silicon nitride films by the reaction of silicon tetrafluoride with ammonia in the temperature range of 1200-1900°C. Other studies using SiF_4/NH_3 have met with limited success.

Silane and Hydrazine Reaction:

A number of investigators have studied the photosensitized decomposition of silane (SiH_4) and hydrazine (N_2H_4) to deposit a silicon nitride layer at relatively low temperatures. Collet (33) used a closed quartz walled system filled with a mixture of approximately equal quantities of silane and hydrazine with a total pressure of 5 torr. Mercury vapor was added to the mixture and was excited with 2537 Å UV light to initiate the decomposition reaction at approx. 200°C. Infrared absorption spectra were similar to pyrolytic nitride material. Dielectric measurements were made on the films and are reported in the dielectric properties section.

Yoshioka & Takayanagi (34) also investigated the silane/hydrazine/hydrogen gaseous mixture reaction over the temperature range between 550 and 1150°C. The hydrazine process gives more rapid deposition rate than the ammonia process. The films produced are transparent, smooth, and tightly adherent. The films from this process have relatively larger etch rates in HF than those from the ammonia process (see etching behavior section). Figures II-21 and II-22 summarize the growth rate data from their experiments.

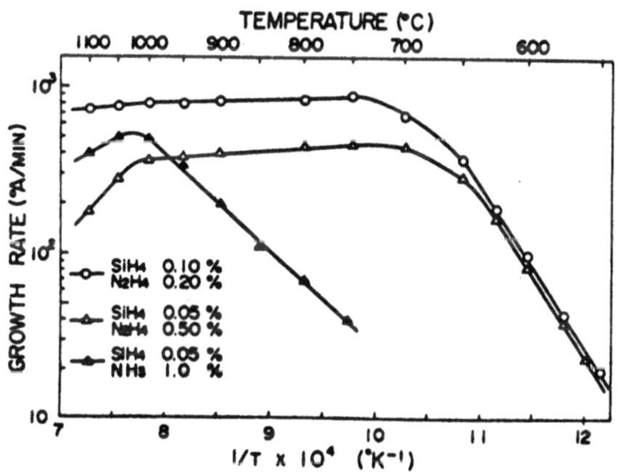

Figure II-21. Growth rate of silicon nitride films vs. substrate temperature for the SiH_4-N_2H_4 process and for the SiH_4-NH_3 process. (34)

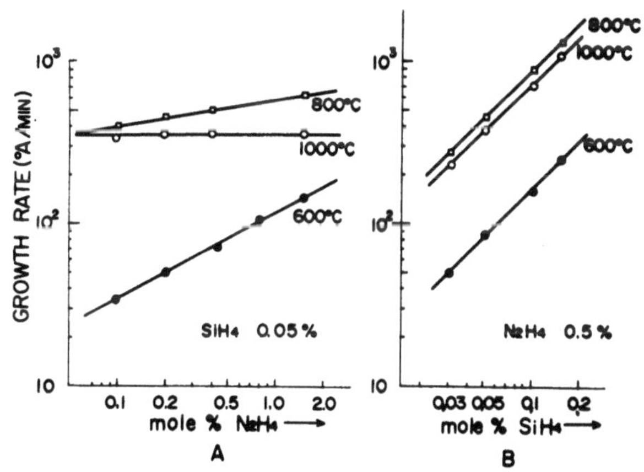

Figure II-22. Dependence of film growth rate. A, mole concentration of N_2H_4 (SiH_4 mole. conc. 0.05%); B, mole concentration of SiH_4 (N_2H_4 mole. conc. 0.5%). (34)

The growth rate reaches a saturation value above 750°C, changes linearly with the silane concentration in the temperature range between 550° and 1000°C, while the dependence on the hydrazine concentration was more complicated becoming gradually independent of the N_2H_4 concentration at high temperatures.

Production Operations

Kohler (4) at Fairchild Semiconductors discussed the use on a production scale of silicon nitride as a passivating layer for a wide range of devices. He noted that the requirements of the process and the film are quite critical. A uniform and reproducible large scale deposition is required (at least ten 2-inch wafers in a run) with good electrical properties in addition to impurity masking. For most devices it is important to keep the electrical properties, such as surface state density and conductivity, under control. Mechanical properties are important to any device because the presence of cracks or pinholes can destroy the masking properties of a good nitride film. Therefore, it is important to determine under which circumstances such defects occur, taking into account the nitride deposition process, the cleaning before deposition, and all subsequent device process conditions. In his study, Kohler compared the silane-ammonia process with the silicon tetrachloride-ammonia process. For the silicon nitride deposition process, a satisfactory compromise must be found to satisfy (1) reproducible and uniform growth rate and density for a large number of wafers in one run, (2) good device passivation with controllable influence of the process on the device characteristics, and (3) conservation of the properties of the silicon nitride after subsequent process steps. Comparative data for the two methods of depositing silicon nitride are given in Table II-3.

Table II-3. Comparison of Silicon Nitride Obtained from SiH_4 and $SiCl_4$ Processes Under Conditions Described for Optimum Processing. (4)

		SiH_4-Silicon Nitride	$SiCl_4$-Silicon Nitride
Structural Quality	Cracking and pinholes Etch rate in 49 pct HF Silicon content	None if film thickness < 2000Å 4Å per sec Stoichiometric	None if film thickness < 1800Å 6Å per sec Stoichiometric
Electrical	Barrier against Na^+ ion contamination at 300°C	v_{FB} shift after heating biasing test ≈ 0.5 v Polarization in the nitride obstructs measurements	v_{FB} shift after heating biasing test <0.5 v
	Same at 535°C	Δv_{FB} 1 to 3 v	Δv_{FB} 1.5 v
	Conductivity (measured at 3×10^6 v per cm)	$\approx 3 \times 10^{-3}$ amp per sq cm $\approx 1.3 \times 10^{-9}$ at 835°C boat temperature	$\sim 6 \times 10^{-10}$ amp per sq cm
	Breakdown voltage	3.9×10^6 v per cm 9.5×10^6 at 835°C boat temperature	8.0×10^6 v per cm
	Fast surface state density at 2000–700Å oxide nitride double layers	2×10^{10} states cm^{-2} (see Ref. 5)	5 to 10×10^{11} states cm^{-2}
Optical	Index of refraction	2.00 to 2.20	1.90 to 1.94

The NH_3-silicon compound gas flow rates (see Pyrolytic Method section for data) used by Kohler were lower than those reported by other laboratories. This low ratio and the low absolute values of the gas flows were necessary to guarantee optimum production processes together with good electrical and mechanical properties.

Cracks and pinholes cannot be tolerated in silicon nitride films produced on a production basis because they provide an opening for impurities and thus defeat the purpose of the film. Kohler (4) notes that such cracking can be caused by the following:

a) high-temperature process cycles after nitride deposition.

b) impurity particles at the surface prior to nitride growth. These can be minimized but not totally avoided in production processing.

c) Oxide steps under the nitride film.

He further states...."It is important to know the maximum nitride thickness that can tolerate normal process conditions and remain free of pinholes and cracks." Thus, in his studies, cracks were first observed at the highest heating temperature. The critical thickness below which no cracking was found after 1200°C heating for 1.5 hours was 2000 Å for silicon nitride from the SiH_4 process, and 1800 Å for the $SiCl_4$ process. Cracking, pinholes, and imperfect areas consisting of impurity particles incorporated in the silicon nitride film were observed in cases where the cleaning prior to deposition was not properly executed and especially in the case where the 10:1 HF etching was omitted. For extremely contaminated interfaces, defects could be observed at thicknesses down to 700 Å. The importance of the 10:1 HF etching is to remove 50 to 100 Å of the oxide surface, thus eliminating most of the impurities embedded in the oxide surface.

In Table II-3, it is evident that the electrical conductivity of the SiH_4 silicon nitride can be decreased by a factor of 10^6 if the boat temperature is raised to 835°C and a higher NH_3 flow rate is used. However, one has to compensate for a low etch rate and fast growth rate by changing the reactant gas ratio and raising the main gas flow rate. Besides the advantage of the low-temperature SiH_4/NH_3 process, which can be preferable for the production of certain silicon devices, another advantage of the 710°C SiH_4 process found by Kohler (4) was the deposition of high-quality silicon nitride over GaAs substrates. A definite disadvantage of the SiH_4 process is the polarization which is observed at all temperature and flow rate ranges.

The Westinghouse Electric Corp. (Aerospace Div., Baltimore, Md.) (35) has prepared a process manual as a guide for depositing silicon nitride films on silicon substrates. The method employed for film formation is the pyrolytic deposition of silane (SiH_4) in an ammonia atmosphere at temperatures of 775 to 975°C. The manual contains both general and specific information with regard to materials, equipment, procedures and process requirements.

Equipment

Since their introduction, the growth of large scale integrated circuits has greatly increased the need for chemical vapor deposition systems as the advantages of deposited dielectric films over thermal or bulk grown films are well known as detailed in other sections of this report. The following is a brief annotation of various commercially available equipment for the preparation of the silicon nitride or combined nitride-oxide structures.

Phoenix Materials Corp. (P.O. Box 9, Kittanning, Pa. 16201) has developed a nitride-oxide deposition reactor, the PMC 700, designed to produce extremely uniform Si_3N_4 or SiO_2 films from silane at the rate of twenty-six (26) 2-inch diameter slices per run. The time required to make a run is approx. 15 minutes for oxide and 25 - 30 minutes for nitride, depending on film thickness. It is equipped for full automatic operation and can also be run manually for various experimental work. The following specifications have been given for the equipment:

Reactor dome.	.18" diameter x 16" high water-cooled stainless
Slice capacity.	.26 two-inch diameter slices
Slice heater.	.14" diameter, resistance type graphite heater
Power supply.	.SCR 15KW at 110 volts, single phase
Temperature Capability.	.300° to 1000°C. ± 1% temperature capability
Run time for Nitride. .	.Approximately twenty-five to thirty minutes
Flow meter array. . . .	.Five streams supplied such as Silane, Oxygen, Nitrogen Ammonia and dopant gas, etc.
Reactor Gas Feed. . . .	.Vertical through rotating feed jet assembly
Reactor Cabinet	.Laminar flow hood 30" x 30" x 72"
Optional.	.Precessing fixture which provides planetary slice action under the rotating gas jet assembly for ultimate in film thickness uniformity - twenty 2" slices.

Applied Materials Technology, Inc. (2999 San Ysidro Way, Santa Clara, Calif. 95051) makes available their AMN-700 series of vapor deposition system to deposit any sequence of silicon dioxide, silicon nitride, and polycrystalline silicon without wafer handling, without removal from the reactor chamber. They claim it to be the first non-rf powered, high-temperature, cold wall reactor for the deposition of dielectric films. It can handle up to twenty 2-inch wafers per station. Typical thickness variation is claimed to be less than ± 5% wafer-to-wafer within a run. Typical properties of Si_3N_4 films deposited with this equipment are etch rate (HF) 4-6 Å/sec; deposition rate 100-500 Å/min; refractive index 2.0; operating temperature range from 650° to 900°C.

METHODS OF PREPARATION-REFERENCES

1. DALTON, J.W. and J. DROBEK. Structure and Sodium Migration in Silicon Nitride Films. ELECTROCHEM. SOC., J., v. 115, no. 8, Aug. 1969. p. 865-868.

2. GRIECO, M.J. et al. Silicon Nitride Thin Films from SiO_4 Plus NH_3-Preparation and Properties. ELECTROCHEM. SOC., J., v. 115, no. 5, May 1968. p. 525-531.

3. ARIZUMI, T. et al. Thermodynamical Analyses and Experiments for the Preparation of Silicon Nitride. JAPAN. J. OF APPL. PHYS., v. 7, no. 9, Sept. 1968.

4. KOHLER, W.A. Structural Properties of Vapor Deposited Silicon Nitride. AIME METALL. SOC., TRANS., v. 246, Mar. 1970. p. 735-740.

5. FRIESER, R.G. Direct Nitridation of Silicon Substrates. ELECTROCHEM. SOC., J., v. 115, no. 10, Oct. 1968. p. 1092-1094.

6. KAISER, W. and C.D. THURMOND. Nitrogen in Silicon. J. OF APPLIED PHYS., v. 30, no. 3, Mar. 1969. p. 427-431.

7. CHU, T.L. et al. The Preparation and Properties of Amorphous Silicon Nitride Films. ELECTROCHEM. SOC., J., v. 114, no. 7, July 1967. p. 717-722.

8. TURKDOGAN, E.T. et al. Silicon Nitrides. Some Physicochemical Properties. J. OF APPLIED CHEM., LONDON, v. 8, 1958. p. 296-302.

9. HECKINGBOTTOM, R. and M.A.G. HALLIWELL. Direct Nitridation of the Si(111) Surface: A Low Energy Electron Diffraction Study. General Post Office, London (Eng.) Research Dept., Rept. no. 61, Dec. 1968. 17 pp. NASA N70-18189.

10. OKADA, K. et al. Properties of Thermally Grown Silicon Nitride Films. PHYS. SOC. OF JAPAN, v. 23, no. 3, Sept. 1967. p. 655.

11. KENDALL, E.J.M. The Stabilization of Silicon Surfaces Using Silicon Nitride. J. PHYS., D, ser. 2, v. 1, 1968. p. 1409-1420.

12. FELDMAN, C. and M. HACSKAYLO. Temperature Characteristics of Vacuum Deposited Dielectric Films. REV. OF SCI. INSTRUMENTS, v. 33, Dec. 1962. p. 1459-1460.

13. ELLIOTT, E. et al. Some Properties of Electron Beam Evaporated Si_3N_4 Films. THIN SOLID FILMS, v. 3, 1969. R47-R48.

14. STERLING, H.F. and R.C.G. SWANN. Chemical Vapour Deposition Promoted by R.F. Discharge. SOLID STATE ELECTRONICS, v. 8, no. 8, Aug. 1965. p. 653-654.

15. SAVAGE, J.A. et al. Investigation of Silicon Nitride Thin Films for Use in Most Devices. Royal Radar Establishment (Malvern, Eng.). RRE Newsletter and Research Rev., no. 7, 1968. p. 15/1-2. NASA N69-28679.

METHODS OF PREPARATION-REFERENCES

16. SWANN, R.C.G. et al. The Preparation and Properties of Thin Film Silicon Nitrogen Compounds Produced by a Radio Frequency Glow Discharge Reaction. ELECTROCHEM. SOC., J., v. 114, no. 7, July 1969. p. 713-717.

17. KUWANO, Y. The Preparation and Properties of Silicon Nitride Produced by a Radio Frequency Glow Discharge Reaction of Silane and Nitrogen. JAPAN. J. OF APPL. PHYS., v. 7, no. 1, Jan. 1968. p. 88.

18. GERETH, R. et al. New Applications for Low-Temperature Ammonia-Free $Si_3 N_4$ Films. Abstract no. 114, Electrochemical Society Meeting, Los Angeles, May 10-15, 1970. p. 293-295.

19. AISENBERG, S. and V. ROHATGI. Study of the Deposition of Single Crystal Silicon, Silicon Dioxide and Silicon Nitride on Cold-Substrate Silicon. Space Sciences Inc., Interim Rept., Apr. 1968. 51 pp. NASA-CR-86101.

20. JANUS, A.R. and G.A. SHIRN. Preparation and Properties of Reactively Sputtered Silicon Nitride. J. OF VACUUM SCI. AND TECHNOL., v. 4, no. 1, Jan./Feb. 1967. p. 37-40.

21. CORDES, L.F. Evidence of Excess Silicon in Reactively Sputtered Silicon Nitride Films. APPLIED PHYS. LETTERS, v. 11, no. 12, Dec. 15, 1967. p. 383-385.

22. HU, S.M. Properties of Amorphous Silicon Nitride Films. ELECTROCHEM. SOC., J., v. 113, no. 7, July 1966. p. 693-698.

23. DOO, V.Y. et al. Preparation and Properties of Pyrolytic Silicon Nitride. ELECTROCHEM. SOC., J., v. 113, no. 12, Dec. 1966. p. 1279-1281.

24. DOO, V.Y. et al. Property Changes in Pyrolytic Silicon Nitride with Reactant Composition Changes. ELECTROCHEM. SOC., J., v. 115, no. 1, Jan. 1968. p. 61-64.

25. DOO. V.Y. and D.R. KERR. Investigation of Refractory Dielectrics for Integrated Circuits. International Business Machines Corp., Mar. 1968. 73 pp. Report no. NASA-CR-995.

26. SUGANO, T. et al. Vapor Deposition of Silicon Nitride Film on Silicon and Properties of MNS Diodes. JAPAN. J. OF APPL. PHYS., v. 7, no. 2, Feb. 1968. p. 122-127.

27. YEARGAN, J.R. and H.L. TAYLOR. Conduction Properties of Pyrolytic Silicon Nitride Films. ELECTROCHEM. SOC., J., v. 115, no. 3, Mar. 1968. p. 273-276.

28. BEAN, K.E. et al. Some Properties of Vapor Deposited Silicon Nitride Films Using the SiH_4-NH_3-H_2 System. ELECTROCHEM. SOC., J., v. 114, no. 7, 1967. p. 733-737.

METHODS OF PREPARATION-REFERENCES

29. SCHAFFER, P.S. and B. SWAROOP. Vapor Phase Growth and DC Breakdown of Silicon Nitride Films. AMERICAN CERAM. SOC., BULL., v. 49, no. 5, 1970. p. 536-538.

30. CARLSON, H.G. et al. Development of Silicon Nitride and Cermet Resistors for Use in a Binary Counter, Metal Insulator Field Effect Transistor Circuit. Texas Instruments Rept. no. 03-68-42, Mar. 1968. 147 pp. NASA N69-10283.

31. DEAL, B.E. et al. Electrical Properties of Vapor-Deposited Silicon Nitride and Silicon Oxide Films on Silicon. ELECTROCHEM, SOC., J., v. 115, no. 3, Mar. 1968. p. 300-307.

32. KUNTZ, V.E. Silicon Nitride Preparation. U.S. Patent 3, 226, 194, Dec. 28, 1965.

33. COLLET, M.G. Depositing Silicon Nitride Layers at Low Temperature Using a Photochemical Reaction. ELECTROCHEM, SOC., J., v. 116, no. 1, Jan. 1969. p. 110-111.

34. YOSHICKA, S. and S. TAKAYANAGI. Deposition of Silicon Nitride Films by the Silane-Hydrazine Process. ELECTROCHEM. SOC., J., v. 114, no. 9, Sept. 1967. p. 962-964.

35. WESTINGHOUSE ELECT. CORP., BALTIMORE, MD. AEROSPACE DIV. Process Manual for Pyrolytic Silicon Nitride Deposition. Contract no. NAS9-6636, 1970. 17 pp. Avail. CFSTI CSCL20L. N70-23899.

PROPERTIES

General

Numerous references have established that the physical, chemical, and electrical properties of deposited dielectric films depend considerably on deposition conditions, e.g., temperature, composition of reactants, etc. Hence, it is important in specifying film property data on silicon nitride to identify the deposition process and conditions; wherever possible, this has been noted in the property data tables in this survey.

The physical and electrical properties of bulk silicon nitride have been well known for some time. It is characterized by high chemical stability and inertness, high electrical resistivity (although lower than SiO_2), high temperature strength and thermal shock resistance, extreme hardness, as well as ability to mask against certain dopants for which SiO_2 is not effective (e.g., Ga, Zn, O_2, and In). N.L. Parr (1) has reviewed the mechanical properties of bulk silicon nitride materials.

In order to fully characterize Si_3N_4 dielectric films on silicon, it is necessary to know the dielectric constant, breakdown field, surface charge, surface recombination velocity, work function difference, conductance, polarization, and trapping behavior. Deal and coworkers (2) at Fairchild Semiconductors have measured and compared silicon nitride films with silicon dioxide films as summarized in Table III-1.

Table III-1. Electrical Properties of Dielectric Films on Silicon (2).

	Thermal SiO_2 (1200°C dry O_2)	Deposited SiO_2 (400°C, SiH_4 + O_2)	Deposited Si_3N_4 (950°C, $SiCl_4$ + NH_3)	Deposited Si_3N_4/thermal SiO_2
Dielectric constant (100 kHz)	3.85	4.52	4.8 – 5.8	—
Breakdown field (d-c)	9.0×10^6 v/cm	8.5×10^6 v/cm	1×10^7 v/cm	—
Surface charge (Q_s')	2×10^{11} cm^{-2}	6×10^{11} cm^{-2}	$>10^{12}$ cm^{-2}	3×10^{11} cm^{-2}
Q_s' orientation dependence	(111) > (100)	None	None	(111) > (100)
Surface recombination velocity (s_o)	5-10 cm/sec	5-10 cm/sec	200 cm/sec	5-10 cm/sec
Work function difference (Al-10^{16} cm^{-3} p-type Si)	−1.0v	−1.0v	−1.0v	−1.0v
Conduction current density at 5×10^6 v/cm (100 v/0.2μ)	8.8×10^{-11} amp/cm^2	2.65×10^{-8} amp/cm^2	1.05×10^{-7} amp/cm^2	—
Polarization	None	Very little	Occurs at 25°C. Increases with temperature and field.	$\propto \frac{x_N}{x_{total}}$
Trapping	None	Occurs at 25°C above critical field.	Occurs at 25°C above critical field.	None if $x_o > 0.1\mu$.

Doo & Kerr (3) have also compared the general properties of SiO_2, Si_3N_4 and $Si\text{-}O_x\text{-}N_y$ summarized in Table III-2.

Table III-2. Summary of the Physical Properties of SiO_2, Si_3N_4 and $Si\text{-}O_x\text{-}N_y$ (3)

Structure	SiO_2 amorphous	Si_3N_4 crystalline	Si_3N_4 amorphous	$Si\text{-}O_x\text{-}N_y$ amorphous
Melting Point (^oC)	~ 1600	~ 1900	-	-
Density (g cm^{-3})	2.2	3.4	3.1	-
Index of Refraction	1.46	2.1	2.05	1.60-1.88
Dielectric Constant	3.8-3.9	9.4	7.5	4.77-6.12
Dielectric Strength (V cm^{-1})	~5 x 10^6	-	~1 x 10^7	~5 x 10^6
Infrared absorption band, (μ)	9.3	10.6	11.5-12.0	9.3 and 12.0
Energy gap (e.v.)	8	3.9-4.0	~5.0	-
Thermal Expansion Coeff. ($^oC^{-1}$)	5.6 x 10^{-7}	3.0 - 3.5 x 10^{-6}	-	-
Thermal Conductivity (cal cm^{-1} sec^{-1} $^oC^{-1}$	0.0032	0.067	-	-
D.C. Resistivity (ohm-cm)				
@ 25°C	10^{14}-10^{16}	10^{15}	~10^{14}	
@ 250°C	-	-		
@ 300°C	-	-		
@ 500°C		~10^{13}	~2 x 10^{13}	
Etch Rate in 10:1 NH_4F:HF (Å/min)	1000	<< 0.1	5-10	33-400

There has been increasing evidence presented in the literature that silicon nitride films can have silicon contamination: Cordes (4), Swann (5), Yeargan & Taylor (6), Coleman & Thomas (7), and Gyulai (8). This contamination can have a pronounced effect on the properties and behavior of the nitride films. The problem lies in the fact that only a few investigators have taken the trouble to analyze for silicon content; if peculiar electrical effects result from the silicon it is oftentime ascribed to other phenomenon because of the lack of specific knowledge regarding the silicon content and behavior within the silicon nitride film layer. Coleman & Thomas (7) have made a serious attempt to develop techniques to identify and measure the silicon by means of electron scattering/radial distribution function technique.

Gyulai (8) correlated electrical and optical property data with silicon content. The results show correlation between deviation from stoichiometry and physical properties of the layers. They used backscattering and channeling effect measurements with MeV ^4He ions to determine the depth dependence of the composition of the amorphous silicon nitride films on single-crystal silicon. The composition was stoichiometric over the entire layer for high ratios of NH_3 to SiH_4 used in the deposition reaction at 850°C. For lower ratios, a silicon excess was found, and in extreme cases, the silicon excess was located predominantly near the interface as shown in Figures III-1 and Table III-3 and tabular data below:

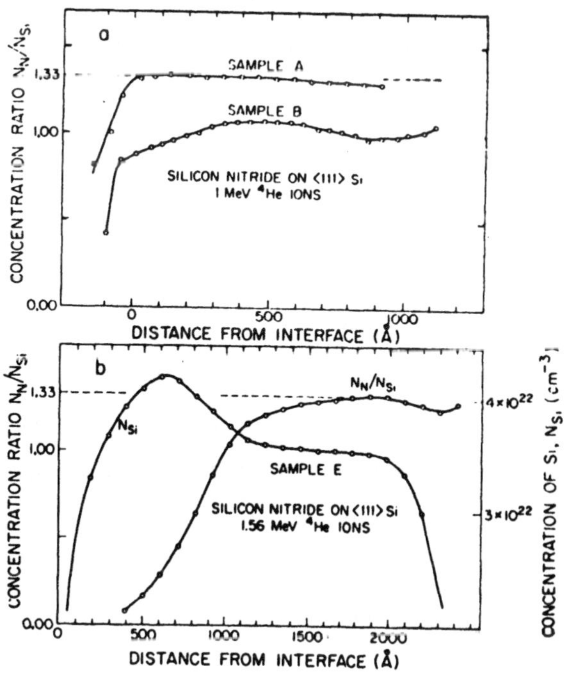

Figure III-1. (a) Concentration ratio of nitrogen to silicon atoms in a silicon nitride layer deposited on a ⟨111⟩ silicon surface. (b) Concentration ratio of nitrogen to silicon atoms in a silicon nitride layer and the concentration of silicon atoms when a layer of excess silicon exists near the interface. (8)

Table III-3. Experimental Data. (8)

Sample	NH_3/SiH_4	Dissolution rate (Å/sec)	N_N/N_{Si} BSM Average	Max value	Thickness (Å) BSM	Ellipsom.	Density* (g cm⁻³) BSM	Diff. wgt. meas.	Dielectric constant	Index of refraction
A	100	5.50	1.31	1.33	929	950	2.82	2.90	7.00	2.00±0.1
B	1	3.00	1.00	1.07	1125±30	1110	2.72	2.95	7.40	2.20±0.1
C	0.25	1.40	0.7	0.75	1115±30	1150	2.42	...	...	2.50±0.2
D	0.125	0.60	0.62	0.69	1135±30	1150	2.53	...	8.70	2.80±0.3

*The BSM densities correspond to the middle portion of the layers.

Crystallography

The crystallography of silicon nitride was first extensively investigated by Forgeng and Decker (9) in 1958 using powder and single crystal samples. The evidence that alpha-silicon nitride (Si_3N_4) forms first on elemental silicon in a nitriding atmosphere was established by a number of experiments carried out by Turkdogan (10) in 1958.

Most investigators of silicon nitride films have reported that the layers are amorphous because x-ray techniques have not indicated any signs of crystallinity, although electron microscopic examination by Hu (11) indicated that small crystalline areas may exist. He notes that the surface texture of films of silicon nitride varies greatly according to the method and conditions of preparation. Foreign impurities and defects on the substrate could result in the formation of crystallites in localized areas according to Chu, Szedon and Lee (12).

Experimental work carried out by Bean, Gleim and Yeakley (13), Cordes (4), Yeargan & Taylor (6) and Deal, Fleming and Castro (2) have established that silicon nitride films contain excess silicon. This had led Kendall (14) to examine nitride films in greater depth. A scanning electron microscope revealed small platelet-type crystalline areas protruding out a little way from the surface in all the films. These were more prevalent in the films grown pyrolytically and in some instances they were up to 10,000 Å across, although in general they were about 100 Å in agreement with other published work (e.g. Dalton and Drobek (15).

Silicon nitride (Si_3N_4) exists in two crystal phases, both having the hexagonal structures although slightly different lattice dimensions. Alpha-silicon nitride is reported to transform to beta-silicon nitride upon heating above 2820°F.

Dalton and Drobek (15) conducted electron diffraction studies on silicon nitride films (prepared by various methods) and found that films varied considerably in crystallite size and structure, alpha, beta, and alpha-beta mixtures of silicon nitride. They observed that pyrolytically deposited (using SiH_4 or $SiCl_4$) films were usually amorphous (< 20 Å); however, if these films were deposited at temperatures < 900°C on improperly cleaned substrates they were crystalline and invariably alpha-silicon nitride. The crystallite size in these crystalline pyrolytic films ranged from 20-40 Å. The structure of Si_3N_4 films prepared by d-c plasma discharge was also studied. Several of them were crystalline; some were amorphous. These crystalline films ranged from 20-80 Å in crystallite size and were usually beta or alpha-beta mixtures. Occasionally, alpha-silicon nitride which forms first and then a transition from alpha to beta was observed at the interface. Microwave plasma deposited films were usually crystalline (alpha or alpha-beta), and average crystallite size ranged from 40 to several hundred angstroms. The diffraction studies also revealed that the films were not uniform in crystallite size throughout their thickness. Some went from amorphous at the top to crystalline at the Si-Si_3N_4 interface, but more often they were amorphous at the interface and crystalline at the top.

Kendall (16) reports from his crystallographic studies and C-V measurements that in view of the similarity of the two hexagonal forms of silicon nitride, that it is relatively easy to change the alpha phase to the beta phase, and the application of a high electric field may be sufficient. Additionally, he reports at room temperature, the two phases may co-inhabit the same crystal, and when there occurs a change from alpha to beta, a stacking fault occurs (which can act as deep-seated electron traps).

<u>Crystal System:</u>	Hexagonal (17)	Hexagonal (18)
<u>Space Group:</u>	α (P31c)	-
	β (Pb 3/m)	-
<u>Lattice Parameters:</u>	α (a = 7.738 Å)	(a = 7.748 ± 0.001 Å)
	α (c = 5.617 Å)	(c = 5.617 ± 0.001 Å)
	β (a = 7.608 Å)	(a = 7.608 ± 0.001 Å)
	β (c - 2.991 Å)	(c = 2.9107±0.0005 Å)

The β-Si_3N_4 is only metastable at 300°K

Chemical Properties

Shaffer (19) has compiled the following chemical property information:

Formula Weight		140.22 g/mole
Theoretical analysis		39.9% nitrogen 60.1% silicon
Resistivity	(Heating in air, 3 hrs at 1100-1400°C)	0.4% weight change
	(Neutral atmospheres)	Useful to 1850°C
	(Reducing atmospheres)	Useful to 1850°C

Etching Behavior

Silicon nitride is a fairly inert chemical compound and the problems inherent in the etching of the material are involved and numerous. There are no fast-acting solvents for the bulk material. Sperry Rand Research Center investigators (20) reported in 1966 that effective etchants were available for the relatively small amount of material in a deposited layer. The etching rate of the nitride is lower than that of the oxide. In silicon nitride etching, the thickness of the film is very important. Numerous investigators have explored the dissolution or etch rates of their silicon nitride films in various chemical solutions: 48% HF (at room temperature), buffered HF (composition: 40g NH_4F + 60ml H_2O + 15ml 48% HF), H_3PO_4 (at 140°-200°C), 70% HF (severe attack), molten NaOH at 450°C, 3 parts HF (49%) + 10 parts HNO_3 (70%) at 70°C.

Van Gelder and Hauser (21) made an extensive study of the etching of silicon nitride in phosphoric acid with silicon dioxide as a mask. They found that the water content of phosphoric acid in etching silicon nitride and silicon dioxide plays an important role. An increase in water content increases the etch rate of silicon nitride and decreases the etch rate of silicon dioxide. Refluxed boiling phosphoric acid at 180°C was found to be a useful etchant for silicon nitride films. The etch rate is 100 Å/min.; under the same conditions, deposited silicon dioxide had an etch rate of 0-25 Å/min. depending on the method of preparation, while elemental silicon etched at 3 Å/min. Figure III-2 summaries the etch rates as a function of temperature.

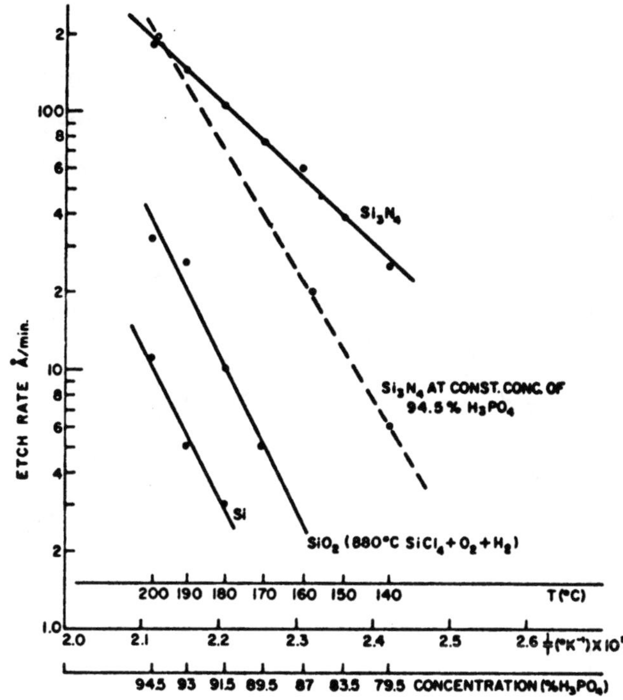

Figure III-2. Solid lines: etch rate of Si_3N_4, SiO_2, and Si in refluxed boiling phosphoric acid at atmospheric pressure as a function of boiling temperature and acid concentration. Dashed line: etch rate of Si_3N_4 at a constant concentration of 94.5% H_3PO_4 as a function of temperature only. (21)

A number of other investigators have utilized HF as an etchant for pyrolytically deposited silicon nitride. Figure III-3 summarizes Doo & Kerr (3) studies using 48% HF on silicon nitride films grown in hydrogen at various silane/ammonia ratios and at different temperatures.

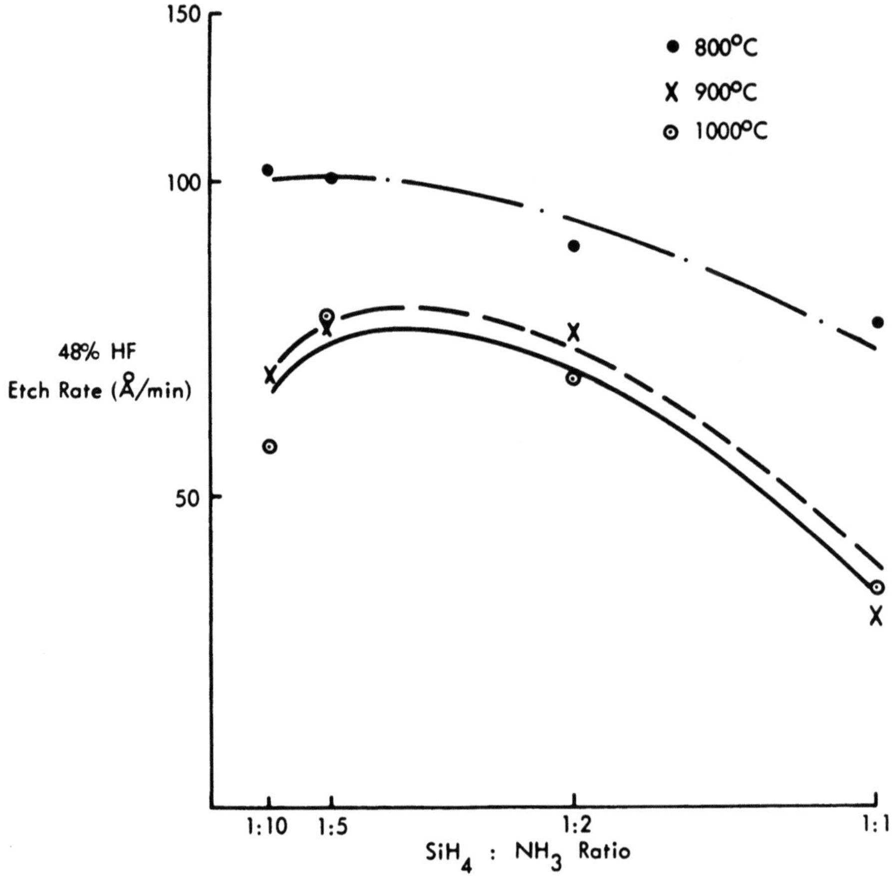

Figure III-3. Etch rate of silicon nitride in 48% HF vs. SiH_4:NH_3 ratio carrier gas H_2 (3).

The etch rate decreases with increasing deposition temperature because the film density should increase with temperature. The 800°C films dissolve much faster than a given temperature decreases with decreasing ammonia injection rate as it approaches the silane injection rate. This could be due to the film composition changing toward the silicon-rich silicon nitride. The etch rate of nitride films grown in nitrogen is plotted as a function of growth temperature in Figure III-4.

Figure III-4. Growth rate vs. temperature and etch rate vs. temperature (3).

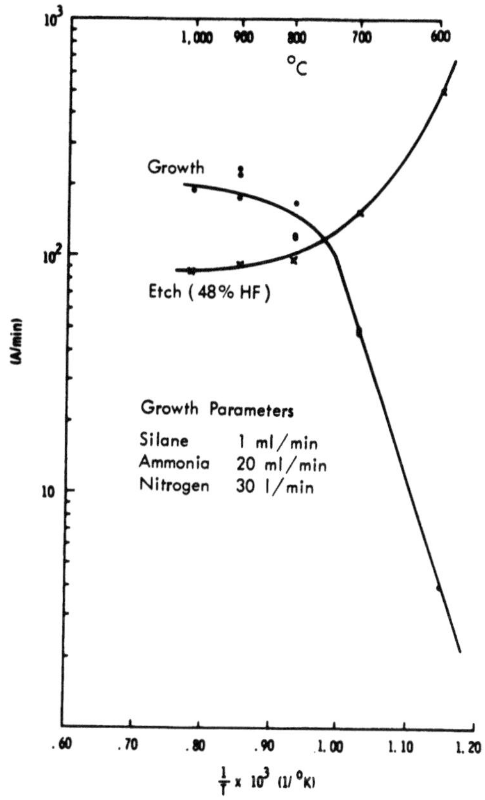

It is apparent that the etch rate increases with decreasing growth temperatures. Chu, Lee and Gruber (22) found similar results for different flow rates as shown in Figure III-5.

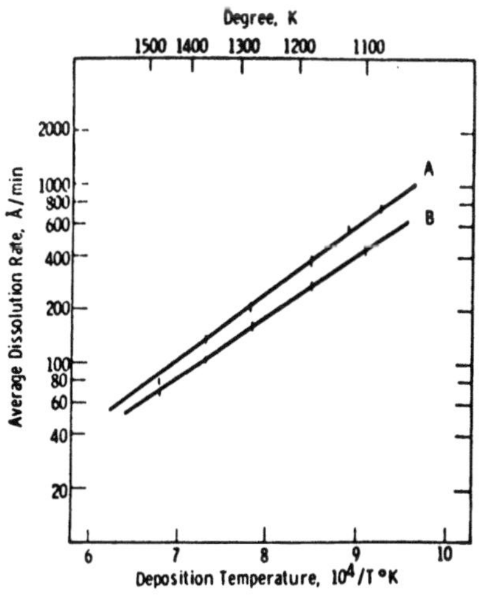

Figure III-5. Dissolution rate of silicon nitride in 49% HF at 24°C as a function of deposition temperature. A: Flow rate of ammonia = 20 l/min, flow rate of SiH_4 = 2ml/min. B: Flow rate of ammonia = 40 l/min, flow rate of SiH_4 = 2 ml/min. (22)

Scott and Olmstead (23) went as low as 600°C with the (SiH_4/NH_3) deposition temperature. They stated that the films deposited between 600 and 800°C were the most etchable while the etch rate decreased as the deposition temperature was increased beyond 800°C. A particularly sharp drop in the etch rate occurred for films deposited or annealed at about 1150°C.

Bean, Gleim and Yeakley (13) found the etch rate increases with etchant temperature for buffered HF (Bell No. 2: 300cc H_2O + 200g NH_4F + 45cc HF) solutions for pyrolytic deposited silicon nitride films as shown in Figure III-6.

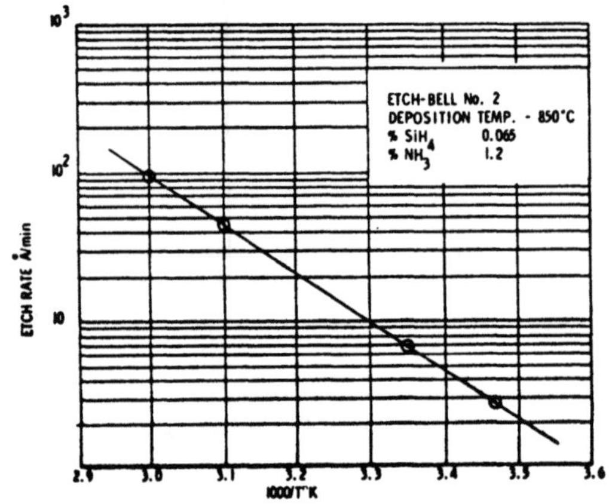

Figure III-6. Etch rate vs. etchant temperature (13).

The etch rate of buffered HF solutions is plotted as a function of percent silane and ammonia in Figures III-7 and III-8.

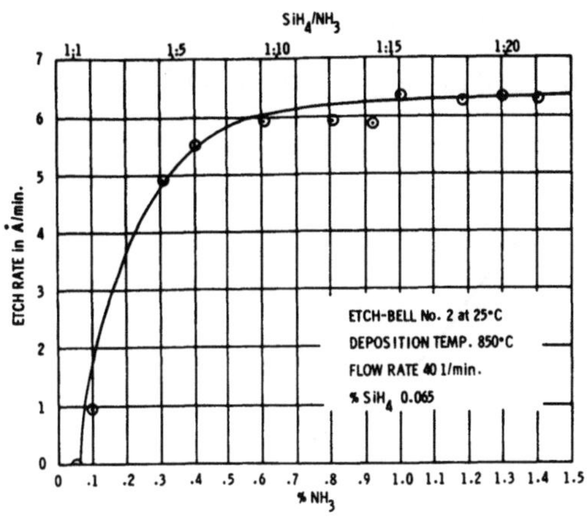

Figure III-7. Etch rate vs. per cent NH_3 (13).

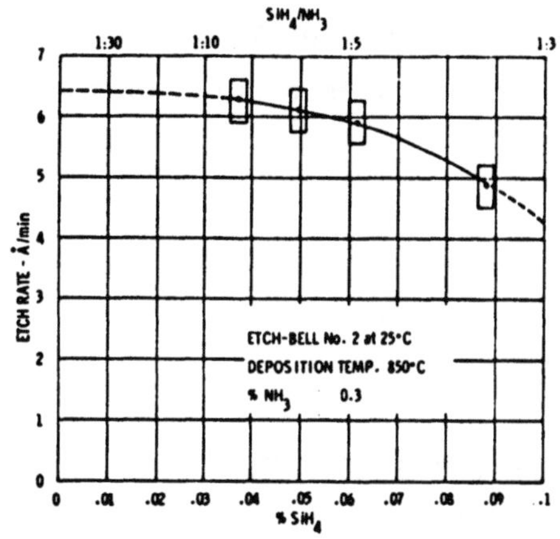

Figure III-8. Etch rate vs. per cent SiH_4 (13).

Thus, it is apparent that above approximately 0.4% NH_3, the etch rate is uniform (approximately 6.25 Å/min), but below this concentration, the etch rate decreases rapidly and approaches zero.

Yoshioka and Takayanagi (24) studied the etching behavior of silicon nitride films prepared pyrolytically by the silane/hydrazine/hydrogen gaseous reaction and found that the films from this process have relatively larger etch rates in HF than those from the ammonia process. Figure III-9 shows the etching rate in 20% HF solution at room temperature to be strikingly dependent on the deposition temperature and higher than those grown at the same temperature by the silane/ammonia process.

Figure III-9. Etch rate of silicon nitride films in a 20% HF vs. deposition temperature in the SiH_4-N_2H_4 process and the SiH_4-NH_3 process. O, SiH_4 0.05%, N_2H_4 0.5%, no heat-treatment after deposition; ●, SiH_4 0.05%, N_2H_4 0.5%, heat-treated for 1 hr at 900°C after deposition; □, SiH_4 0.05%, NH_3 1.0%, no heat-treatment after deposition. (24)

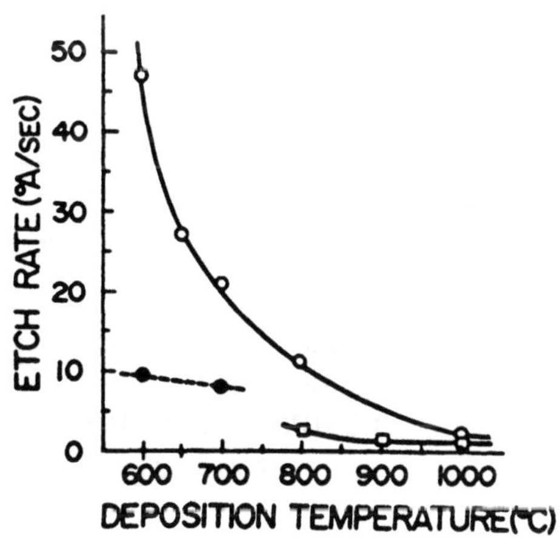

Figure III-10 provides Kohler's (25) data on the etch rate variation with relative gas flow rates of silicon nitride films made by the $SiCl_4$-NH_3-H_2 pyrolytic process.

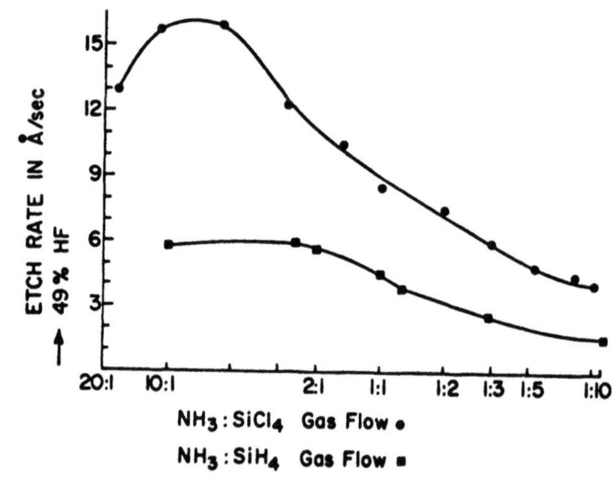

Figure III-10. Etch rate in 49% HF vs NH_3:$SiCl_4$ gas flow ratios. (25)

A comparison of the etch rates of the SiH_4/NH_3 processes as a function of heating time and temperature is shown in Figures III-11 and III-12.

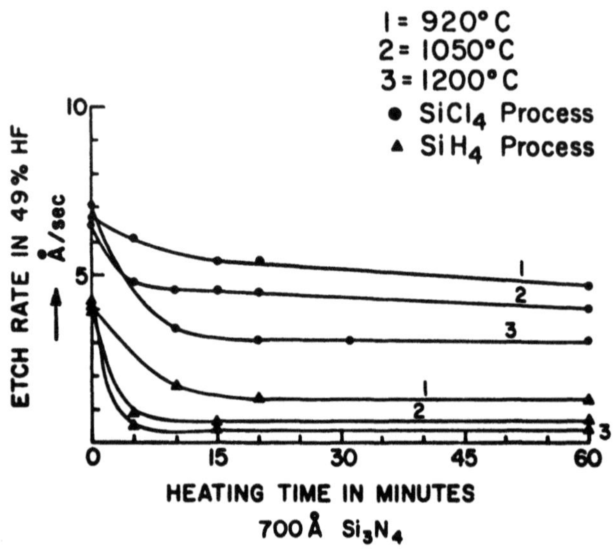

Figure III-11. Silicon nitride etch rate in 49% HF vs heating time. (25)

Figure III-12. Silicon nitride etch rate in 49% HF vs heating time at 1200°C. (25)

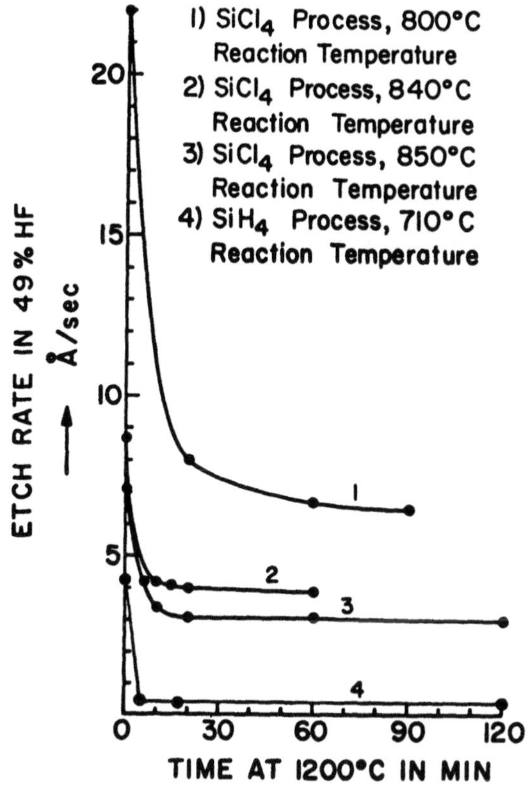

Sterling & Swann (26) in their glow discharge method studies on silicon nitride films reported that films formed in the cold are soluble in hydrofluoric acid but when formed at 300°C are resistant to acid attack. If such films are either laid down or subsequently heat treated at 700-900°C, they become extremely resistant chemically and appear to be unaffected by normal etches, including CP4. Joyce, Sterling and Alexander (27) also investigated the chemical resistance of glow discharge prepared films and found the chemical resistance to be dependent on the conditions of deposition as shown in Figure III-13. They believe that the conditions of radio frequency coupling affect the type of deposit and can change its chemical properties. Films which can be etched easily at 1000 Å/sec, and are deposited in a capacitive discharge, contrast with films which are more chemically resistant and are deposited in an inductive type of discharge.

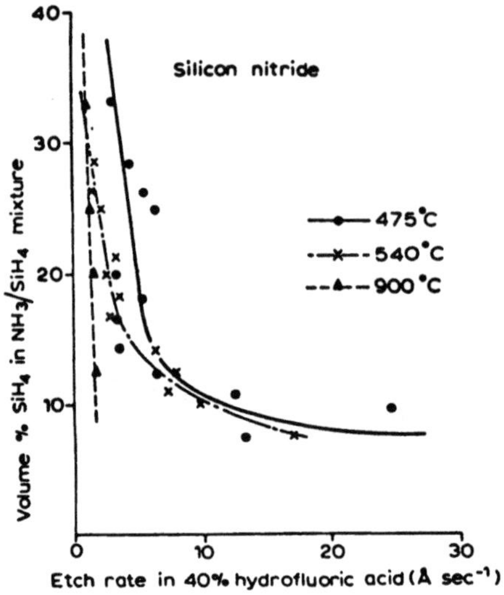

Figure III-13. Etch rate of silicon nitride. (27)

Hu & Gregor (28) and Gregor (17) have prepared reactively sputtered silicon nitride and measured the etch rates of such films. They found a considerable amount of scatter because of the dependence on substrate temperature, power density, film thickness, and gas pressure when using 7:1 buffered HF solutions.

Diode system films (i.e. D-C reactive sputtered Si_3N_4) have been found to etch rapidly by Janus & Shirn (29). P-etch rates at room temperature for triode systems films are given in Figures III-14 and III-15 as a function of time and sputtering voltage:

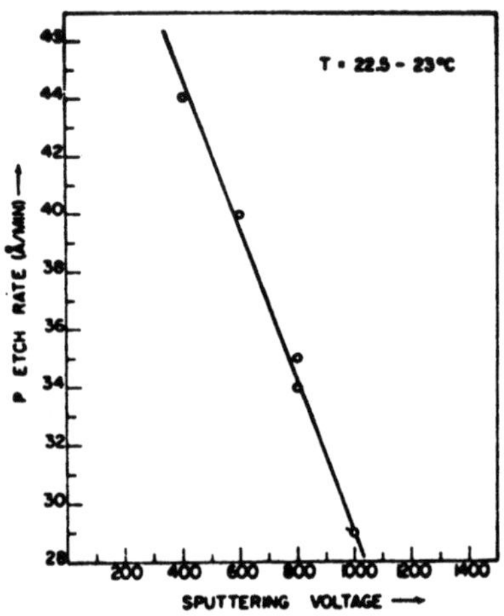

Figure III-14. P-etch rate dependence on sputtering voltage. (29)

Figure III-15. P-etch rate for reactively sputtered silicon nitride. (29)

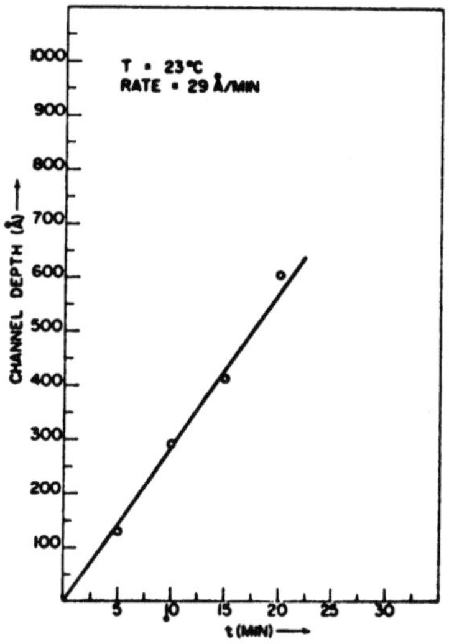

Films etched in 48% HF had reported rates of 120 Å/min, which increased (upon a 1100°C heat treatment in vacuum at 1 x 10^{-6} torr and also in N_2 and N_2-Ar atmospheres) to 150 Å/min.

The Transene Company Inc. (30) has developed and sells commercially TRANSETCH-N which selectively etches silicon nitride in the presence of silicon dioxide or silicon. It produces openings in the deposit of silicon nitride films with a degree of resolution comparable to the best obtainable with silicon dioxide in conventional planar technology, according to the manufacturer. It contains no fluoride, produces no undercutting, provides a wide margin of safety in selection of etching times. The product is a purified reagent derived from ortho-phosphoric acid, colorless, odorless liquid, boiling point 180°C, and is ready to use as supplied.

Silicon nitride films on silicon substrates can be also converted to SiO_2 by anodic or electrochemical oxidation under various current density conditions in a suitable solvent electrolytic combination according to experiments conducted by Tripp (31) and Schmidt and Wonsidler (32). Solvent solutions used in these studies included (a) a 7.5 v/o solution of pyrophosphoric acid in tetrahydrofurfuryl alcohol, (b) ethylene glycol + 2% water, 0.4M in KNO_3. Figures III-16, III-17 and III-18 and Table III-4 summarize the film conversion process variables and effects:

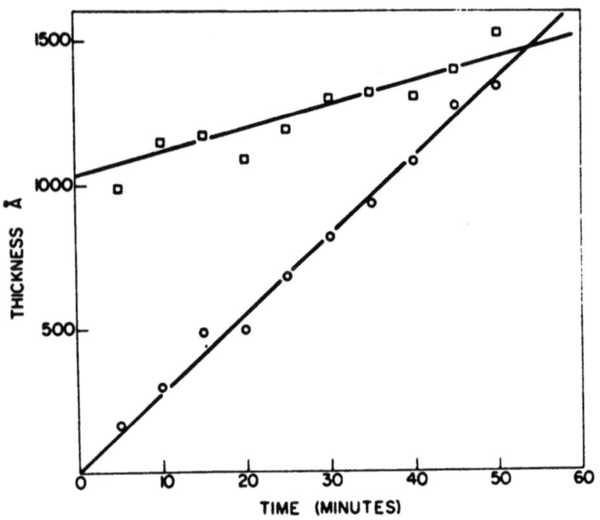

Figure III-16. Total film thickness and converted film thickness as a function of time during the anodic oxidation of silicon nitride films on silicon at a current density of 7.5 mA/cm^2.
□ Total film thickness;
○ converted film thickness.
(31)

Current densities, mA/cm²	Total film growth rate, Å/min	Conversion rate, Å converted film formed/min	Normalized conversion rate (1/i dD/dt), (Å converted film formed × cm²)/(min × mA)
4.0	3.93	13.1	3.28
5.5	5.50	18.9	3.44
7.5	8.20	27.4	3.65
9.0	10.5	35.0	3.89
11.0	12.0	46.4	4.22
13.0	8.0	56.3	4.33

Table III-4. Conversion rate, total film growth, and normalized conversion rate for anodization of silicon nitride films on silicon at various current densities. (31)

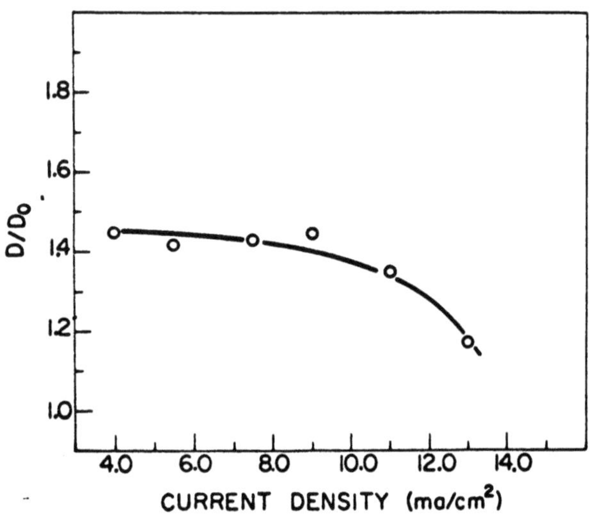

Figure III-17. Ratio of completely converted film thickness to initial silicon nitride film thickness as a function of current density. (31)

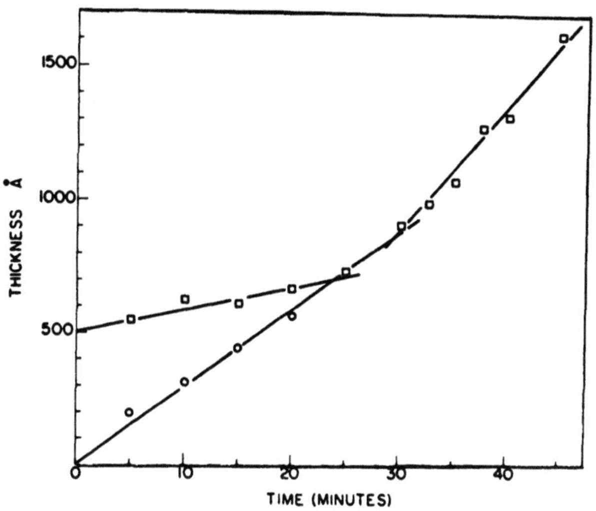

Figure III-18. Total film thickness and converted film thickness as a function of time during the anodic oxidation of 500 Å silicon nitride films on silicon at a current density of 7.5 mA/cm^2. □ Total film thickness; ○ converted film thickness. (31)

From these data, it is apparent that the rate of conversion of silicon nitride to silicon dioxide is independent of time and that the conversion efficiency increases with increasing current density.

Brown, Engeler and Garfinkel (33) developed a new masking technique using molybdenum as an integral film for etching very high resolution patterns in passivating films of silicon nitride. The Mo film has much better masking properties than the photoresists used on Si_3N_4 previously. Table III-5 compares the etch rates of Si_3N_4 with SiO_2 and silicon oxy-nitride films. Table III-6 lists the sequence of process steps in the Mo mask technique.

Table III-5. Film Etch Rates. (33)

	In concentrated HF, A/min	In buffered HF, A/min
Silicon dioxide	50,000	1000
Silicon nitride*	150	15
Silicon oxy-nitride**	3,000	125

Diffusion mask	Time to etch through mask In: Buffered HF, min	Conc. HF	KPR Photoresist Life† In: Buffered HF, min	Conc. HF, min
5000A SiO_2	5	6 sec	10	—
1000A Si_3N_4	66	7 min	—	<½

* Si_3N_4 films referred to here have been grown by the pyrolytic decomposition of silane (SiH_4) and ammonia (NH_3) gases at 1000°C. The etching properties of silicon nitride depend on the conditions for its formation and on subsequent heat-treatments.
** The silicon-oxy-nitride compound is prepared in the same manner as the silicon nitride except in this instance small amounts of O_2 are added to the ammonia and silane gas mixture.
† Photoresist life in the etching solution is defined as the approximate time it takes to lift the edges of the photoresist mask or crack the photoresist mask or for the etch to permeate the photoresist to an extent that destroys the resolution of the pattern.

Table III-6. Mo Mask Process Steps. (33)

1.	Silicon nitride coating	Any thickness without cracks and pinholes is workable.
2.	Mo deposit	Evaporated or sputtered* Mo films. About 1000A or thicker is suitable, depending on the desired resolution.
3.	Photoresist	Standard KPR techniques.
4.	Mo etching	A few seconds in ferricyanide etch, depending on thickness. The etch rate of the Mo film in this etch is about 150A/sec.
5.	Silicon nitride etching	Concentrated (48%) HF.
6.	KPR and Mo removal	Hot cleaning solution, or other methods.

* Electron beam evaporation in an ion pumped oil free vacuum system and "Triode" sputtering in Ar gas was used in this investigation.

According to Feist (34), pure and dense Si_3N_4 etches at a rather slow rate (the etch rate of amorphous Si_3N_4 in 48% HF is approximately 100 Å/min). This causes difficulties with conventional photomasking because the latter is destroyed by the etchant during the necessary long etching time. Although techniques using more durable masking (e.g. Cr or SiO_2, permitting the use of hot H_3PO_4 as an etchant) have been worked out, the alternative of using composites of Si_3N_4 and SiO_2 has also received attention. These silicon oxynitrides etch faster than Si_3N_4, while retaining much of its favorable masking and passivating properties.

Optical Properties

Refractive Index

The index of refraction of a dielectric film is to a certain extent indicative of the density (or porosity) and its composition. The various methods for measuring and reporting refractive index data on silicon nitride have been used. Caution must be practiced with regard to wavelengths used and thinness of samples. The substrate temperature and deposition temperatures influenced the refractive index properties according to Chu, Lee and Gruber (22). Table III-7 summarizes the various values at room temperature reported in the literature for the refractive index of silicon nitride.

TABLE III-7. Refractive Index of Silicon Nitride Films at Room Temperature.

Deposition Method	Film Type	Wavelength	Refractive Index	Reference
Bulk Samples	Crystalline		2.10	Chu (22) Janus & Shirn (29)
Film		(Wedge fringe method)	2.02 ± 0.09	Janus & Shirn (29)
Film		(Oil immersion method)	2.08 ± 0.02	Janus & Shirn (29)
Film			2.00 ± 0.02	Chu (22) Grieco (35)
Vapor Deposited		5460 Å	1.96	Deal (2)
Reactively Sputtered	Amorphous		2	Cervanak (36)
Silicon Nitridation	1670 Å Thick		2.65	Okada (37)
Silicon Nitridation	1530 Å Thick		2.80	Okada (37)
Silicon Nitridation	1340 Å Thick		3.10	Okada (37)
Pyrolytic (SiH_4/N_2H_4)	Amorphous		2.0 - 2.1	Yoshioka & Takayanagi (24)
Glow Discharge	(0.1% SiH_4)		1.95	Gereth (38)
	(13% SiH_4)		2.9	Gereth (38)

The index of refraction of silicon nitride films as determined from ellipsometer measurements at 5461 Å as a function of ammonia percentage (silane reaction) are shown in Figure III-19.

Figure III-19. Index of refraction vs. percent NH_3. (13)

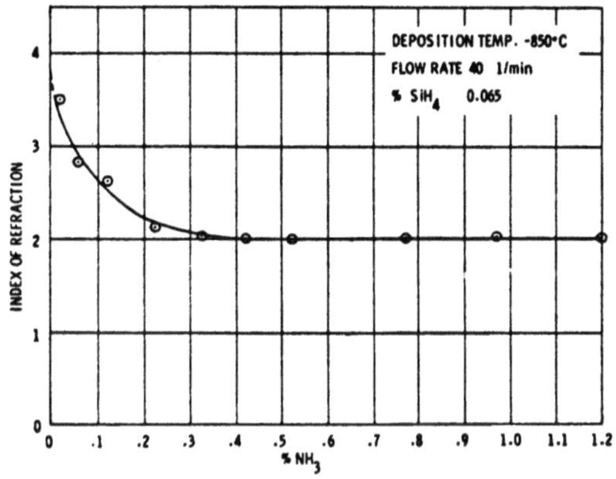

It is apparent that over most of the compositional range studied, the index of refraction varied from 2.0 to 2.05, but as the stream percentage of ammonia decreased below 0.3%, an increase in the index was noted (silicon-rich nature of such films). Kohler (25) found that the index can vary with gas flow rate in the SiH_4/NH_3 process as shown in Figures III-20 and III-21.

Figure III-20. Index of refraction of silicon nitride as a function of the NH_3 gas flow in the SiH_4 system. (25)

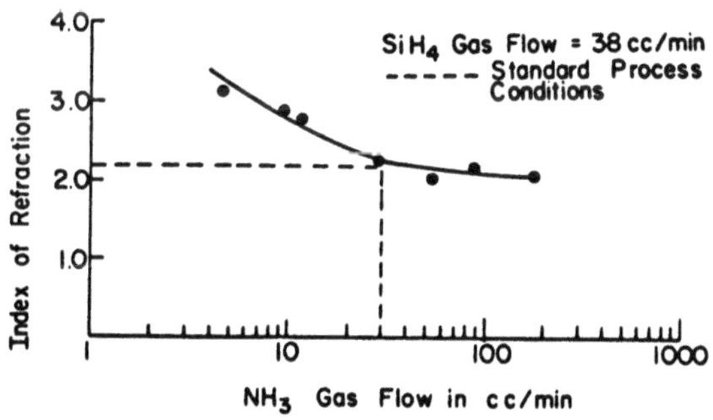

Figure III-21. Index of refraction of silicon nitride along the susceptor in the $SiCl_4$ system. (25)

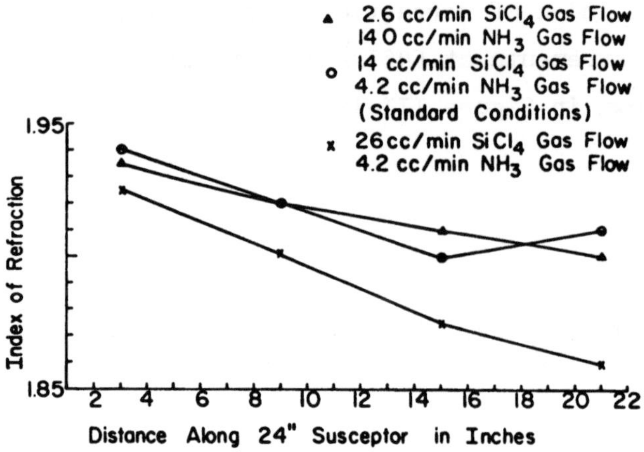

Figure III-22 provides Reizman's and Van Gelder's (39) data on the refractive index vs wavelength for silicon nitride films in the visible and a comparison with SiO_2 films.

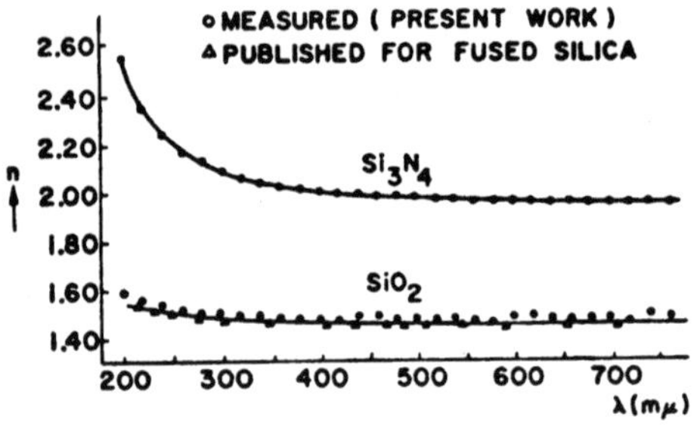

Figure III-22. Refractive index vs. wavelength for SiO_2 and Si_3N_4. (39)

Figure III-23 provides the refractive index of silicon nitride films grown in hydrogen carrier gas as a function of silane to ammonia ratio.

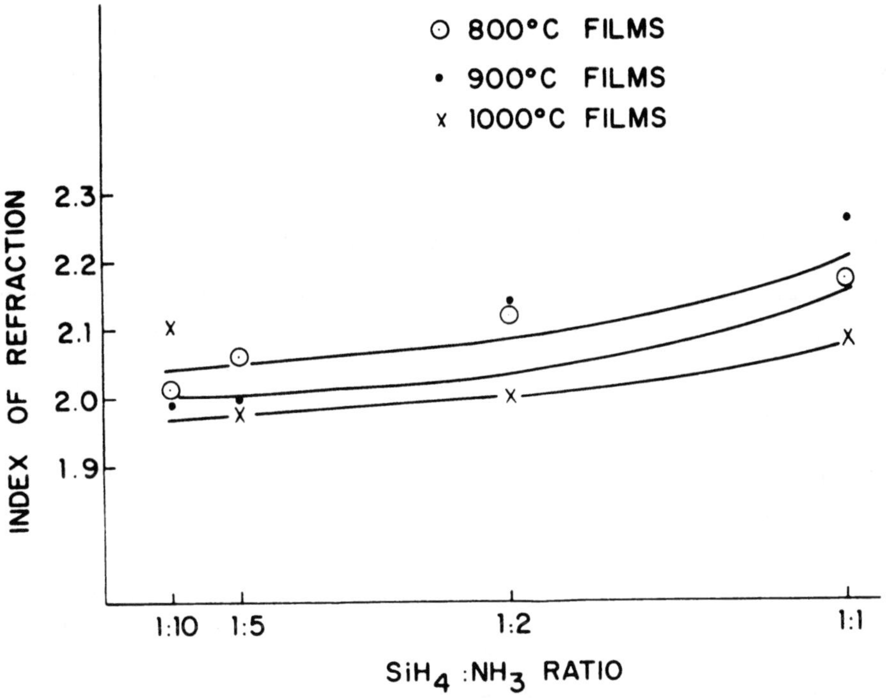

Figure III-23. Refractive index of silicon nitride vs. $SiH_4:NH_3$ ratio hydrogen grown films. (3)

In general, the refractive index increases with decreasing ammonia injection rates at all the deposition temperatures (from 800° to 1000°C). When the ammonia injection rate equals the silane injection rate, the refractive index of the amorphous silicon nitride films exceeds that of the crystalline alpha-silicon nitride (n = 2.1). The refractive index is also temperature dependent-decreases with temperature. Similar results were obtained with films grown in nitrogen carrier gas.

Color/Thickness Relationship

A color comparison is given in Table III-8 for silicon nitride and silicon dioxide films as a function of film thickness.

Table III-8. Color Comparison of SiO_2 and Si_3N_4 Films (39).

Order	Color	SiO_2 Thickness Range[a] (μ m)	Si_3N_4 Thickness Range (μ m)
	Silicon	0-0.027	0-0.020
	Brown	0.027-0.053	0.020-0.040
	Golden Brown	0.053-0.073	0.040-0.055
	Red	0.073-0.097	0.055-0.073
	Deep Blue	0.097-0.10	0.073-0.077
First	Blue	0.10-0.12	0.077-0.093
	Pale Blue	0.12-0.13	0.093-0.10
	Very Pale Blue	0.13-0.15	0.10-0.11
	Silicon	0.15-0.16	0.11-0.12
	Light Yellow	0.16-0.17	0.12-0.13
	Yellow	0.17-0.20	0.13-0.15
	Orange-Red	0.20-0.24	0.15-0.18
First	Red	0.24-0.25	0.18-0.19
	Dark Red	0.25-0.28	0.19-0.21
Second	Blue	0.28-0.31	0.21-0.23
	Blue-Green	0.31-0.33	0.23-0.25
	Light Green	0.33-0.37	0.25-0.28
	Orange-Yellow	0.37-0.40	0.28-0.30
Second	Red	0.40-0.44	0.30-0.33

[a] $n(Si_3N_4)/n(SiO_2) = 1.97/1.48 = 1.33$

Spectra

Optical transmittance data on silicon nitride films over the range from 0.2 to 24 microns are given in Figures III-24 and III-25.

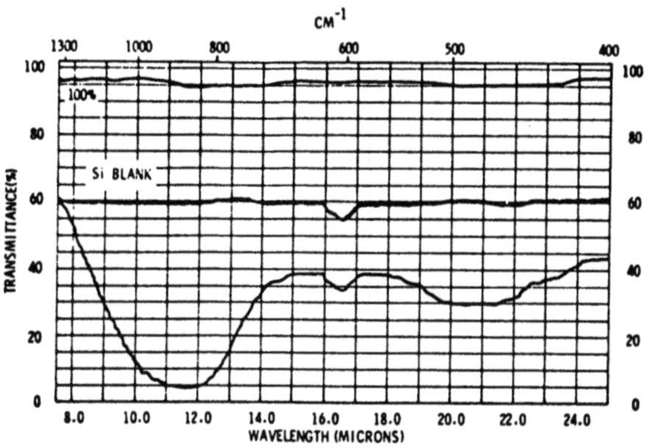

Figure III-24. Infrared transmission (8.0-24.0 μ) of silicon nitride. (13)

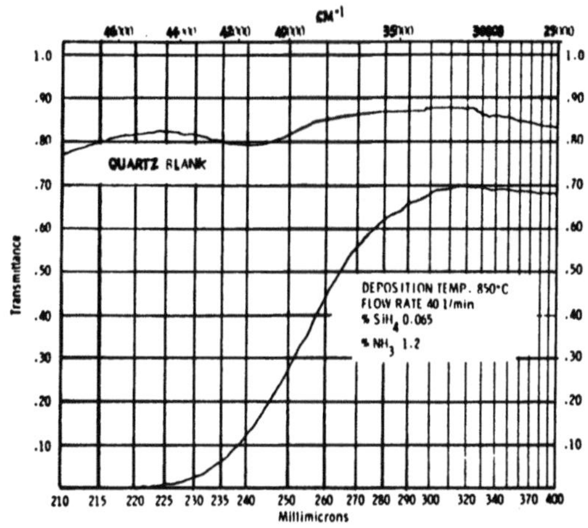

Figure III-25. Optical transmission (0.2-0.4 μ) of silicon nitride. (13)

Between 0.4 and 8 microns, there appear to be no absorption bands; above 8 microns the most prominent absorption is that due to the Si-N bond which occurs in the 10-12 micron range.

Mechanical Properties

Hardness

Microhardness measurements were made on a series of 10,000 Å thick silicon nitride films deposited on a silicon substrate by Bean, Gleim and Yeakley (13). The silane concentration was held constant at 0.065% with NH_3 percentage varying from 0.05 to 1.2. Their results are shown in Figure III-26:

Figure III-26. Knoop hardness vs. percent NH_3. (13)

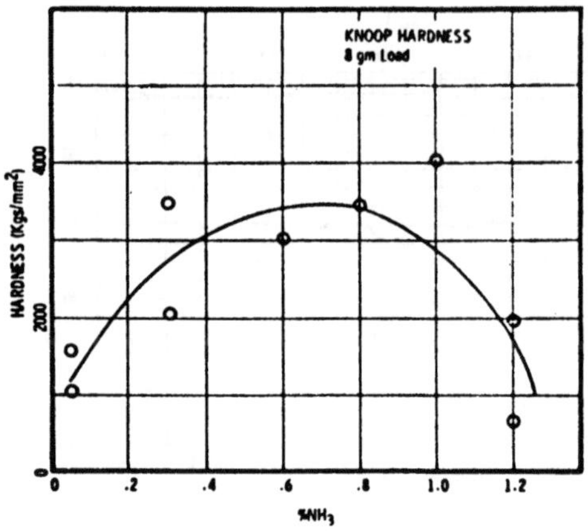

It is apparent that softer films occur at each end of the composition range. Moh's hardness measurements were also made as given in Figure III-27 on a series of films deposited at various temperatures from 800° to 1200°C.

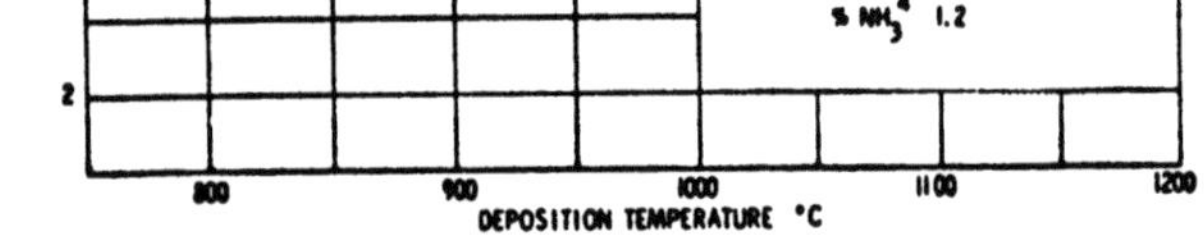

Figure III-27. Moh's hardness vs. deposition temperature. (13)

Other Properties

Silicon-silicon nitride structures differ from silicon-silica structures in that chemically deposited silicon nitride films are in tension while silica films are under compression. It is expected that films of silica-silicon nitride mixtures of proper composition deposited on silicon would have very low stress. Drum and Rand (40) conducted stress studies of these films and means to overcome these stress problems which can cause film cracking, elastic distortion of thin substrates, and plastic deformations in the silicon. They have pointed out that SiO_2 and Si_3N_4 films exhibit stresses of opposite signs (41). Therefore, by depositing a mixture of the two materials in the form of a silicon oxynitride, the stress is zero for films of the composition $Si_3O_4N_2$.

Other mechanical property data compiled from the literature are given in Table III-9.

Table III-9. Miscellaneous Mechanical Properties of Silicon Nitride Films.

Property	Film	Data	Reference
Young's Modulus	(Si_3N_4 film)*	9.6×10^6 psi	Bean (13)
	(bulk crystal)	13×10^6 psi	Shaffer (19)
	(pyrolytic SiH_4/NH_3)	4.0×10^{10} g/cm	Tokuyama(42)
Breaking Strength	(Si_3N_4 film)*	67,000 psi	Bean et al (13)
Tensile Strength	(bulk crystal)	15,800 psi	Shaffer (19)

* Values obtained from one film 0.34 mil thick deposited from a gas composition of 0.1% NH_3, 0.065% SiH_4, and 99.84% H_2.

Physical Properties

Density

Chu, Lee and Gruber (21) reported on the densities of their amorphous silicon nitride films prepared by the techniques shown in Table III-10:

Table III-10. Density of Silicon Nitride Films. (22)

Reaction	Deposition temp. °C	Density, g cm^{-3}
$SiH_4 + NH_3$	850	2.78
	950	2.82
	1100	2.92
$SiCl_4 + NH_3$	1200	3.01
α-Si_3N_4, theoretical		3.18

The density of the films are dependent on the substrate and deposition temperature.

Table III-11 summarizes density data obtained by other investigators.

Table III-11. Silicon Nitride Film Densities.

Sample	Crystal Identification	Density	Reference
Commercial Si_3N_4		3.44	Doo (43)
Pyrolytic (SiH_4/NH_3)		3.02-3.21	Doo (43)
Pyrolytic (SiH_4/N_2H_4)	(dep. at 600-1000°C)	3.0-3.1	Yoshioka & Takayanagi (24)
Pyrolytic (SiH_4/NH_3)	(dep. at 1000°C)	3.1	Philipp (44)

Thermal Properties

One of the thermal advantages of silicon nitride is that its coefficient of thermal expansion is closer to that of silicon than is that of the oxide, thus less mechanical strain should occur at the interface of a MIS structure. Silicon nitride is also a good conductor of heat. Table III-12 summarizes the available thermal property data.

Table III-12. Thermal Expansion of Silicon Nitride.

Property	Material Form	Property Data	Reference
Thermal Conductivity	Crystal (Bulk)	0.067 cal/cm/sec/°C	Doo (3)
Specific Heat	Crystal (Bulk)	0.17 cal/gm/°C	Shaffer (19)
Thermal Expansion	Crystal (Bulk)		
	α Form	3.0×10^{-6}/°C	Doo (3)
	α Form	2.11×10^{-6} per °C; 25-500°C	Shaffer (19)
		2.87×10^{-6} per °C; 25-1000°C	
		3.66×10^{-6} per °C; 25-1500°C	
	β Form	1.47×10^{-6} per °C; 25-500°C	Shaffer (19)
		2.25×10^{-6} per °C; 25-1000°C	
		3.66×10^{-6} per °C; 25-1500°C	
		2.5×10^{-6} per °C	
	Pyrolytic Deposit ($SiH_4/NH_3/H_2$ system)	4.2×10^{-6} per °C; 0-1000°C	Bean (13)
	Pyrolytic Deposit (SiH_4/NH_3 system, 800-1000°C)	3.85×10^{-6} per °C (25-1000°C)	Tokuyama (42)

Burkhardt and Marvel (45) measured the thermal expansion of sputtered silicon nitride films and reported the data in Figure III-28 as compared with that of silicon.

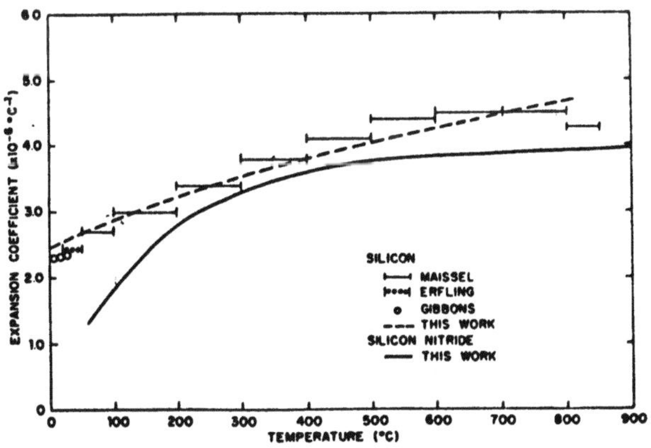

Figure III-28. Variation, with temperature, of the coefficient of thermal expansion of silicon and of a silicon nitride film. (45)

Electrical Properties

It is unfortunate that silicon nitride does not quite match the favorable characteristics of SiO_2, due to its smaller band gap (crystalline Si_3N_4: E_g = 3.9 - 4 eV, amorphous Si_3N_4: $E_g \sim 5$ eV, vs amorphous SiO_2: $E_g \sim 8$ eV). In addition, the relatively low energy barriers towards silicon or metals which Si_3N_4 appears to form, facilitate the injection of electrons or holes into the dielectric at fields $E \gtrsim 2 \times 10^6$ V/cm. A consequence of this charge injection is the pronounced hysteresis appearing in the capacitance-voltage curve of metal-insulator-semiconductor structures after the voltage has been swept to large values. For these reasons, it is now customary either to form a thin SiO_2 barrier layer ($\sim$ 200 Å or thicker) on the silicon surface before the Si_3N_4 layer is deposited, or to form most of the dielectric of SiO_2, and to seal its surface with a protecting layer of Si_3N_4 (a few hundred angstroms thick), according to Feist (34).

Dielectric Constant

The reported values of the dielectric constant for silicon nitride films range from 4 to 12.7, apparently depending on the deposition method. In general, film dielectric properties are comparable to bulk values, however, no general agreement on the bulk dielectric constant value is evident as indicated in Table III-12.

Table III-12. Bulk Silicon Nitride Dielectric Constant.

Value	Temp.	Frequency	Density	Reference
9.4	not stated	not given	not stated	Janus & Shirn (29)
9.4	not given	not given	not stated	Popper (46)
5.5±0.9	room temp.	8.2-12.4 KMc	2.37 g/cc	Wells (47)
6.3±0.5	room temp.	8.2-12.4 KMc	2.30 g/cc	Wells (47)

Table III-13 summarizes the various dielectric constant measurements of films as reported in the literature.

Table III-13. Dielectric Constants for Silicon Nitride Films.

Deposition Method	Substrate Deposition Temperature	Measurement Frequency	Measurement Temperature	ε'	Notes	Reference
r.f. Glow Discharge	200°C	1 kHz	20°C	6.2		Sterling & Swan (26)
SiH_4/NH_3	400°C	1 kHz	20°C	9.4		
Glow Discharge Sputtering (reactive) $3Si + 2N_2$	50 - 370°C	-	-	6.2-6.8	2.82-3.02 ±0.14 Film Density	Janus & Shirn (29)
Pyrolytic SiH_4/NH_3	750 - 1100°C	-	-	6.34 ±0.08	3.02-3.21 Film Density	Doo et al (43)
Pyrolytic	-	10 kHz	-	6.2	Amorphous Films	Hu (11)
Reactively Sputtered	-	10 kHz	-	6.5	Amorphous Films	Hu (11)
Pyrolytic (uv light) Photosensitized SiH_4/N_2H_4	200°C	-	-	5.5		Collet (48)
Pyrolytic SiH_4/NH_3	900°C	-	-	6.0		Collet (48)
Reactively Sputtered $3Si + 2N_2$	500 - 600°C	2 kHz-10 kHz		6 - 7	Amorphous	Cervenak (36)

Table III-13. Dielectric Constants for Silicon Nitride Films. (Continued)

Deposition Method	Substrate Deposition Temperature	Measurement Frequency	Measurement Temperature	ε '	Notes	Reference
Vacuum Deposited	-	1 kHz	-	4.2	1μ Thick	Feldman & Hocskaylo (49)
Pyrolysis $SiBr_4 + N_2$	960°C	1 kHz	25°C 600°C	12.7 13.5	Mo Substrate	Barnes & Geesner (50)
Glow Discharge $3Si + 2N_2$	500 - 600°C	1 kHz	-	6-7	Annealed Film treatment in 10^{-6} Torr vacuum	Aleksandrov (51)
Pyrolytic $SiCl_4 + NH_3$	950°C	100 kHz	-	4.8-5.8		Deal (2)
-	-	-	-	7.5	Amorphous	Doo et al (3)
Pyrolytic SiH_4/N_2H_4	800°C	-	-	6.1	3.0-3.1 g/cm^3 Film Density	Yoshioka & Takayanagi (24)
Glow Discharge $SiH_4 + N_2$ SiH_4 (0.1-5%)	-	-	-	4-7		Kuwano (52)

Table III-13. Dielectric Constants for Silicon Nitride Films. (Continued)

Deposition Method	Substrate Deposition Temperature	Measurement Frequency	Measurement Temperature	ε'	Notes	Reference
Pyrolytic	850°C	10 kHz	20°C	7.5-8.7	1000-1300 Å thick, homogeneous; some crystallinity in films	Kendall (16,53)
Glow Discharge SiH_4/NH_3	350°C	10 kHz	20°C	7.5-8.7		

The Melpar Co. in 1962 measured the dielectric constant as a function of thickness for sputtered films and obtained an average value of 4.2 when evaporated at 2100°C and the substrate temperature 250°C (The films were deposited at 5×10^{-5} torr at a rate of about 300 Å/min). Janus and Shirn (29) have also reported dielectric constant measurements on reactively sputtered silicon nitride films having different bulk values as plotted in Figure III-29.

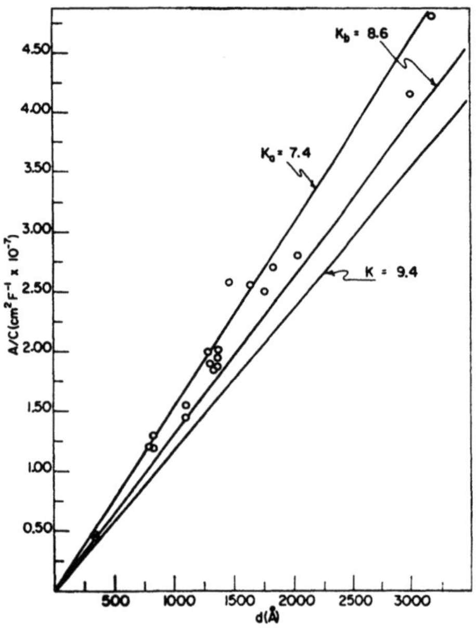

Figure III-29. Dielectric constant plot for reactively sputtered silicon nitride. (29)

They noted that the dielectric constant for the film material is apparently less than the bulk value.

Swann (5) in glow discharge reaction studies on preparing Si_3N_4 films measured the dielectric constant as shown in Figures III-30 and III-31 as a function of silane concentration for film thicknesses of approximately 1000 Å. Its value is seen to change from 7 to 11 for a silane concentration range from 3 to 50%.

The dielectric constant of pyrolytic silicon nitride (amorphous) films made through the gas phase reaction of silane-ammonia procedure as determined by Doo (54) is given in Figure III-32.

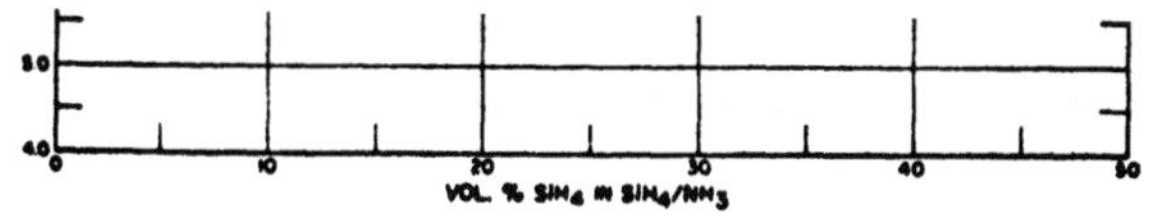

Figure III-30. Dielectric constant vs. silane concentration (5).

Figure III-31. Dielectric constant vs. film thickness (5).

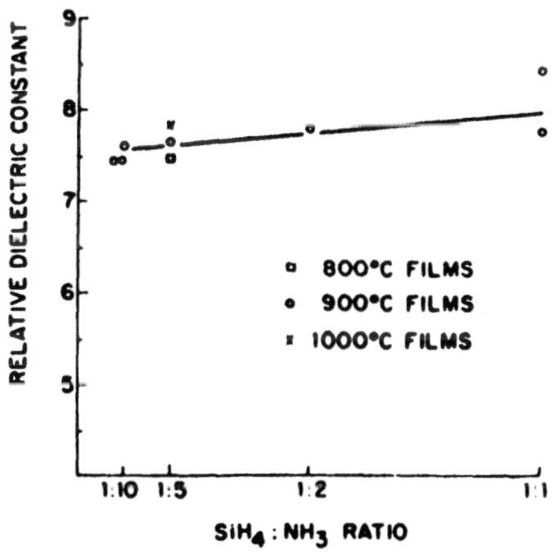

Figure III-32. Relative dielectric constant of pyrolytic silicon nitride films. Measurements are at 25°C and 10 kHz. Each data point represents an individual wafer. □, 800°C films; O, 900°C films; X, 1000°C films. H_2 = carrier gas (54).

From these data, it is evident that the relative dielectric constant has very little dependence on either deposition temperature or the $SiH_4:HN_3$ ratio. Doo and Kerr concluded also from other experiments that the dielectric constant is not sensitive to the carrier gas used (3).

Dissipation Factor

The loss factor of films deposited by the glow discharge reaction method (SiH_4 + NH_3) as a function of silane concentration shows a trend of becoming more lossy as amorphous silicon is approached. A typical value at 1 kc/s is 0.007 increasing to 0.002 at 50% concentration (5).

The dissipation factor (at 1 kc) of silicon nitride films produced pyrolytically from $SiBr_4$ + N_2 + H_2 at 960°C from 25°C to 600°C are given in Table III-13.

Table III-13. Dissipation Factor for Silicon Nitride Films (50).

Temp. °C	Dissipation factor at 1 kc
25	<0.0001
100	<0.0001
200	<0.0001
300	<0.0001
400	0.0003
500	0.0006
600	0.0033

Electrical Resistivity

The electrical resistivity and breakdown of silicon nitride films as a function of silane concentration in the glow discharge method studies as determined by Swann (5) is given in Figure III-33.

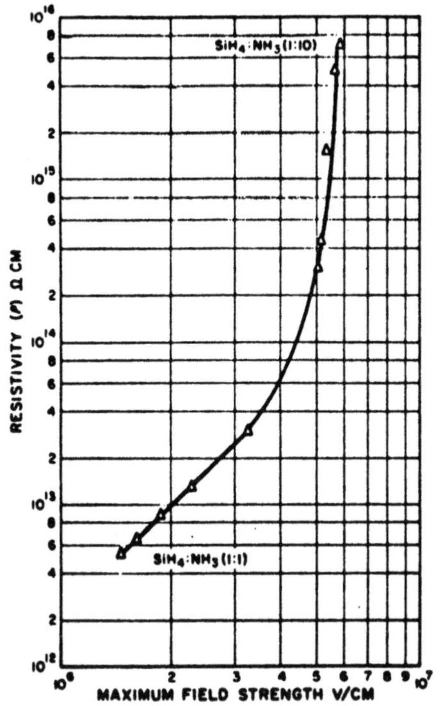

Figure III-33. Resistivity vs. dielectric field strength (5).

Table III-14 summarizes available electrical resistivity data for silicon nitride films.

Table III-14. Electrical Resistivity of Silicon Nitride Films

Deposition Method	Deposition Temperature	Electrical Resistivity Ω-cm	Temperature	Notes	Reference
		∿ 1×10^{15} 2×10^{13}	400°C 500°C		Hu (11)
Pyrolytic $SiBr_4/N_2H_2$	960°C	4.1×10^{16} 1.6×10^{13}	25°C 600°C		Barnes & Geesner (50)
		$>10^{14}$		MNS diode structure used	Neuman (20)
		10^{11}		Stain fibers on Si	Yamazaki (55)
		10^{15} ∿ 10^{13}	25°C 350°C	Crystalline Si_3N_4	Doo & Kerr (3)
		∿ 10^{14} ∿ 2×10^{13}	25°C 350°C	Amorphous Si_3N_4	
		5×10^{14}	25°C		Tombs (56)
Pyrolytic Glow Discharge SiH_4/NH_3	350°C	5×10^{14}	25°C	1150 Å thick, homogeneous	Kendall (16)

Table III-15. Dielectric Strength of Silicon Nitride Films.

Deposition Method	Deposition Temperature	Film Density	Dielectric Strength	Notes	Reference
Glow Discharge $3\ Si + 2N_2$	500-600°C		5×10^5-5×10^6 V/cm		Aleksandrov (51)
			$\sim 1 \times 10^7$ V/cm	amorphous	Doo & Kerr (3)
Pyrolytic SiH_4/N_2H_4	600-1000°C	3.0-3.1 g/cm^3	6-10 $\times 10^6$ V/cm		Yoshioka and Takayanagi (24)
Glow Discharge $SiH_4 + N_2$	260°C		5×10^6 V/cm		Kuwano (52)
Glow Discharge Pyrolytic SiH_4/NH_3	850°C 350°C		10^7 V/cm	1000-1300 Å thick, homogeneous some crystallinity in film	Kendall (16) (53)
Electron Beam Evaporation	250°C		5×10^5-10^6 V/cm	Al-Si_3N_4-Au structure	Elliott (57)
Glow Discharge 0.1% SiH_4 13% SiH_4	<400°C		10^7 V/cm 5×10^6 V/cm		Gereth (38)

Deposition Method	Deposition Temperature	Film Density	Dielectric Strength	Notes	Reference
Pyrolytic $SiCl_4 + NH_3$	950°C		1×10^7 V/cm	D-C breakdown method	Deal (2)
rf Glow Discharge SiH_4/NH_3			$> 5 \times 10^6$ V/cm	1 micron thick film	Sterling and Swann (26)
Pyrolytic SiH_4/NH_3	750°-1100°C	3.02-3.21	$\sim 10^7$ V/cm		Doo (43)
- Sputtered - Pyrolytic			$\sim 1 \times 10^7$ V/cm	amorphous films	Hu (11)
Reactive Sputtering			2×10^5- 7×10^6 V/cm	amorphous films	Cervenak (36)
Pyrolytic SiH_4/N_2N_4	200°C		10^6 V/cm	uv light sensitized reaction	Collet (48)
Pyrolytic SiH_4/NH_3	900°C		10^7 V/cm		
Pyrolytic $SiCl_4/NH_3$			$0.5\text{-}1 \times 10^7$ V/cm	amorphous films	Grieco (35)
Pyrolytic $SiBr_4/N_2/H_2$	960°C		1000 V/mil 600 V/mil	(25°C) (600°C)	Barnes and Geesner (50)
			$\sim 10^7$ V/cm	strain films on Si	Yamazaki (55)

Dielectric Strength

The various dielectric strengths obtained on silicon nitride films produced by the different deposition processes are compiled in Table III-15. In general, the values fall between the range 10^5 to 10^7 V/cm. Data reported by Schaffer and Swaroop (58) on breakdown voltages for various thicknesses of silicon nitride and silicon dioxide-nitride films produced by the chemical vapor deposition technique (SiH_4 + NH_3 + H_2) are given in Table III-16. Their data indicate the average dielectric strength of nitride films is of the order of 10^7 V/cm.

Additional voltage breakdown data are reported in the I-V Characteristics section of this chapter with a brief discussion of the current transport behavior.

The relationship between the dielectric strength and resistivity for silicon nitride films prepared by the glow discharge method are shown in Figure III-34. The breakdown field strength decreased from approx. 6 x 10^6 V/cm at 9% silane to 1 x 10^6 V/cm at 50% for these films.

Table III-16. Breakdown Voltage for Various Thicknesses of Silicon Nitride and Silicon Dioxide-Nitride Films (58)

Specimen No.	Substrate	Thickness (A) Oxide	Thickness (A) Nitride	Deposition temp. (Optical pyrometer) (°C)	Maximum breakdown voltage Si(−)	Maximum breakdown voltage Si(+)
1	*b*		5000	950	530	574
2	*a*		5000	950	580	658
3	*a*		5000	1000	917	1000
4	*b*		5000	1000	640	680
5	*a*		5000	1100	658	430
6	*b*	1200(*v*)	2500	1000	353	530
7	*a*	1200(*v*)	4000	1000	505	528
8	*b*	1200(*t*)	2500	1000	503	585
9	*a*	1200(*t*)	3000	1000	802	863
10	*a*	1200(*t*)	5300	1000	192	245
11	*a*	1200(*t*)	4000	900	411	523
12	*a*	1200(*t*)	4000	950	545	650
13	*a*	1200(*t*)	4000	1000	970	1020
14	*a*	1200(*t*)	4000	1100	1050	1100

NOTE: *b* = Silicon doped with 10^{18} atoms antimony/cm^3 silicon
a = Silicon doped with 5×10^{18} atoms antimony/cm^3 silicon
v = Vapor-deposited oxide
t = Thermal oxide

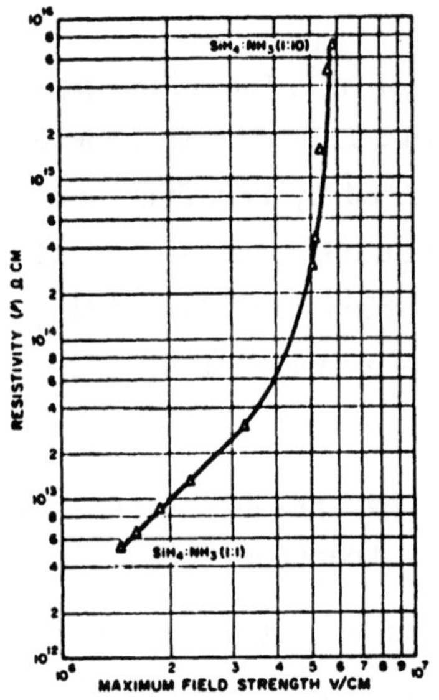

Figure III-34. Resistivity vs. dielectric field strength (5).

Sze's (59) data are given in Figure III-35 for the behavior of the maximum dielectric strength as a function of silicon nitride films prepared by the pyrolytic reaction of $SiCl_4$ and NH_3 at 1000°C and deposited on a degenerate silicon substrate (0.0005 ohm-cm, n-type).

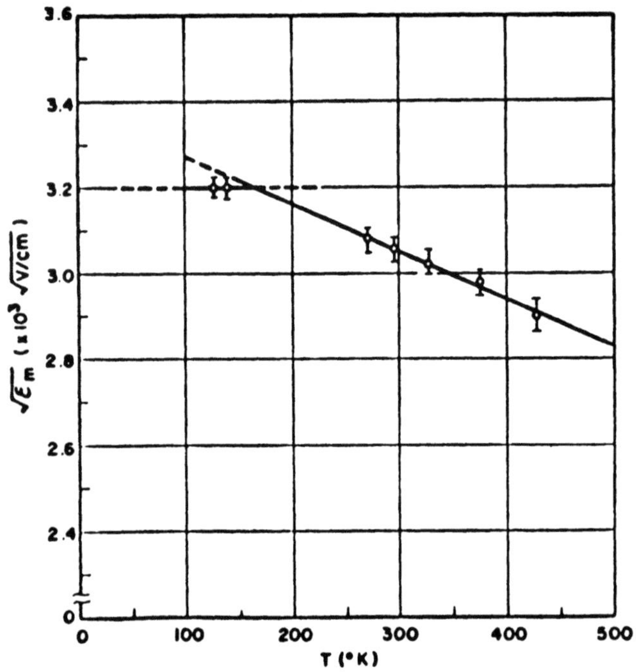

Figure III-35. Square root of maximum dielectric strength versus temperature of silicon nitride films (59).

Energy Gap

Kendall (16) in studying deep-seated electron traps and their electrical behavior postulates that when alpha-and beta-forms of silicon nitride exist in a crystal, stacking faults can occur and possibly act as deep-seated electron traps.
Figure III-36 provides a possible band structure that may exist along the [0001] axis of the crystalline areas of the nitride.

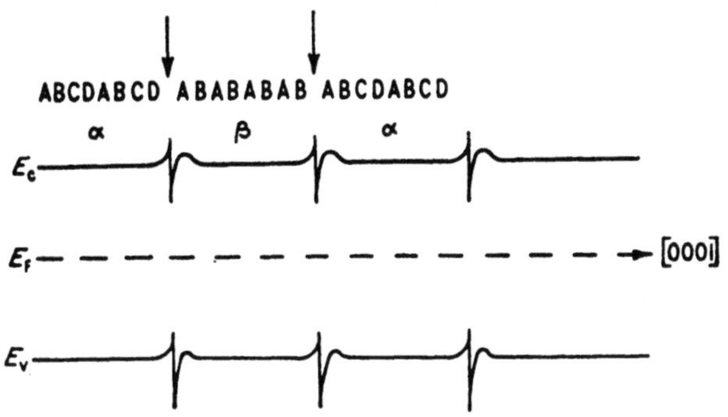

Figure III-36. A possible band structure for the crystalline Si_3N_4 (16).

The similarity (hexagonal) of the two structures could account for a relatively easy change from the alpha phase to the beta phase, and the application of a high electric field may be sufficient to accomplish this. Once the field is removed, the nitride film would return to its equilibrium state, the newly-formed beta reverting to the alpha phase. Thermodynamical considerations of the alpha-to-beta transition indicates that this mechanism is kinetically possible.

Kendall's model is offered to explain the "easy-shift" and "hard-shift" regions along the bias axis. As the traps are filling (or emptying), the characteristic will move quickly along the axis until they are full. If a high electric field is maintained there is a slow inversion of the alpha-Si_3N_4 to extra beta-Si_3N_4, thus increasing the trap density. The new traps fill immediately. When the applied field is removed, the new traps disappear, the extra beta-Si_3N_4 reverting back to alpha-Si_3N_4.

Beta-Si_3N_4 is only metastable at 300°K; it may not exist at all at -78°K, thus no stacking faults exist and no deep-seated traps would arise in this temperature regime. C-V curve data at -78°K (liquid nitrogen temperatures) bear out the above reasoning by Kendall.

A band picture for silicon nitride proposed by Yeargan (60) is presented in Figure III-37.

Figure III-37. Band Structure Proposed by Yeargan (20)

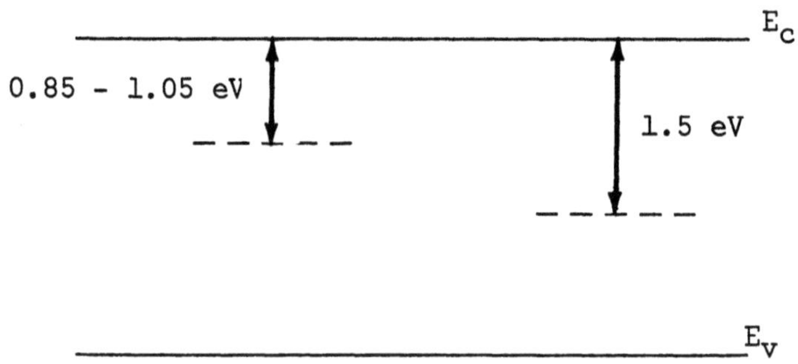

At approximately 1.5 eV below the conduction band there is a trap level corresponding to the intrinsic properties of silicon nitride. At levels varying from 0.85 to 1.05 eV there exist a series of trap levels arising from small islands of silicon embedded in the Si_3N_4. The conduction behavior of Si_3N_4 films on silicon and quartz substrates with different contacts was also investigated. Measurements on substrates of these materials with film thicknesses ranging from approximately 80 Å to 2500 Å established the major current injection mechanism as thermal emission from the bulk of the film. Breakdown strengths of the films were approximately 1×10^7 V/cm.

In 1969, Kendall (14) at the University of Birmingham reported that the energies of his traps were found to vary between 0.50 and 0.90 eV below the conduction band of the nitride using thermally stimulated current techniques. Present evidence suggests that the distribution observed is due to shallow nitride states and interface states about 0.10 eV below the conduction band of the silicon, and probably connected with the trap-hopping process which takes place at around 100°K and has an activation energy of 0.10 eV. As a result of his studies, he agrees with the model proposed above by Yeargan (20) but submits a slight modification as shown in Figure III-38.

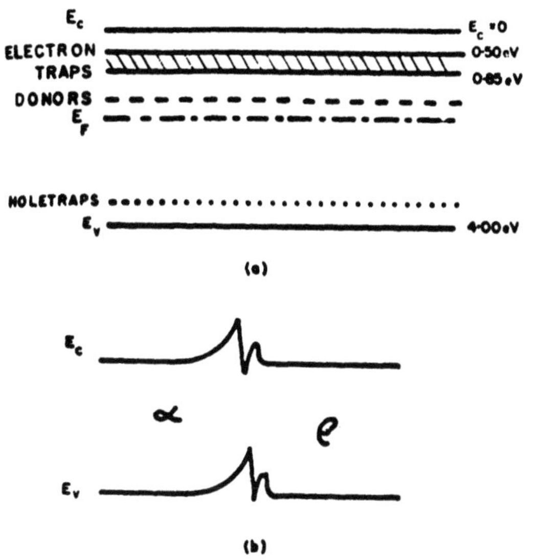

Figure III-38. A possible simplified band structure for Si_3N_4 as a whole (a) and showing the anomalies giving rise to structural type traps (b). (53)

I-V Characteristics & Current Transport

Sze (59) has reviewed and studied the current transport and maximum dielectric strength of silicon nitride films under various voltage and temperature conditions in order to obtain an understanding of the conduction processes operative in these films. Current-voltage curves are given in Figures III-39 and III-40 for silicon nitride films (1100 Å thick) deposited on degenerate silicon substrates (0.0005 ohm-cm, n-type) by the process of reacting $SiCl_4$ and NH_3 at 1000°C.

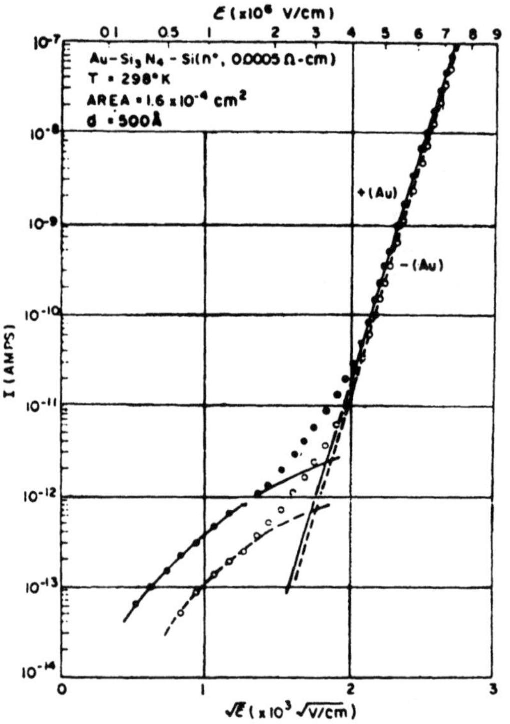

Figure III-39. Current-voltage characteristics of Au-Si_3N_4-Si diode at room temperature (59).

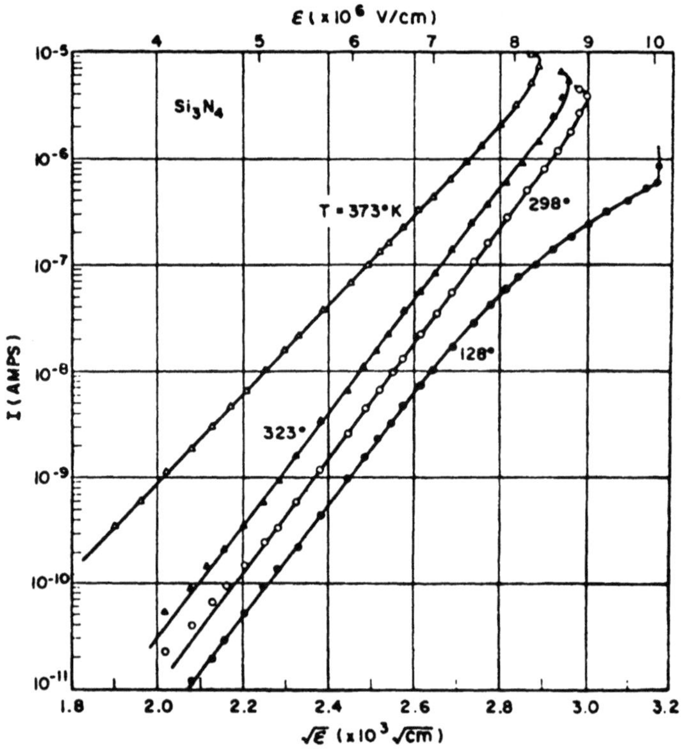

Figure III-40. Current-voltage curves of a typical nitride film 1100 Å thick and 0.2 mm in diameter at four different ambient temperatures. (59)

Tests made on silicon nitride films deposited at 900°C with helium carrier gas and varied $SiH_4:NH_3$ ratios with the result that the log I vs $E^{1/2}$ slopes were similar to those observed with hydrogen carrier gas. Table III-17 summarizes these experiments.

Table III-17. Effect of $SiH_4:NH_3$ on Electronic Conduction of Nitride Films Deposited at 900°C in He Carrier. (3),

Wafer	$SiH_4:NH_3$	Field for I/A = 10^{-6} A/cm^2 (10^6V/cm)
4284	1:1	±5.6
4285	1:5	±5.3
4286	1:10	±5.0
4287	1:20	±4.7

Comparison with Figure III-40 (H_2 carrier gas) indicates the effect of varying the $SiH_4:NH_3$ ratio is suprisingly small.

Kendall (53) has made I-V measurements at room temperature on silicon nitride films and which show that the nitride does not exhibit a simple Frenkel-Poole effect but has three bulk-limited components. At about 20 volts applied bias, a sharp change occurs in the characteristic which represents the threshold field required to fill the deep-seated traps that have been found to exist in the nitride.

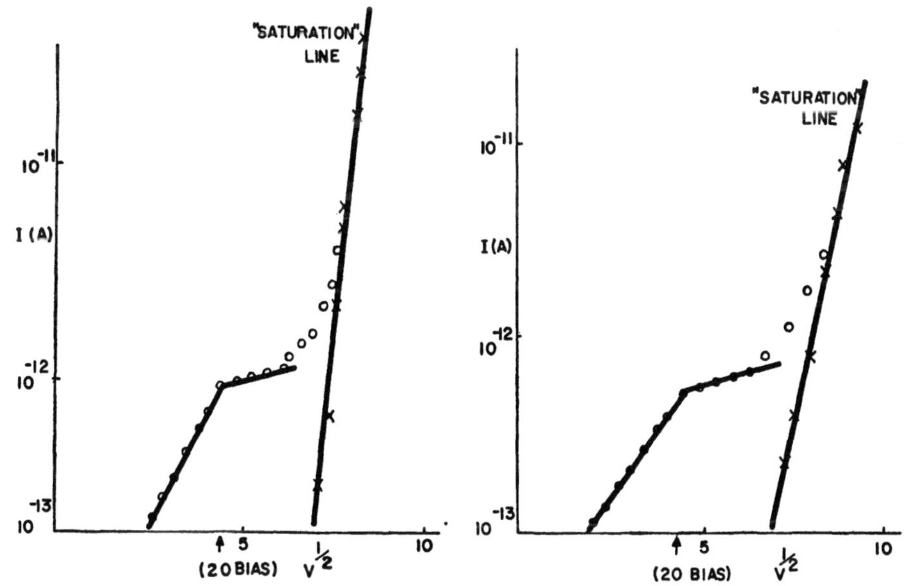

Figure III-41. I-V characteristics at 300°K for two typical devices. (53)

He is of the opinion that.....

> "So far current-voltage results indicate a Frenkel-Poole mechanism for charge transfer to and from both types of trap. Current-temperature measurements also indicate some field ionization and trap hopping." (16)

He concludes that the C-V method, and to a lesser extent the G-V method, does not yield reliable information about the surface states. Both methods only give the value of the effective concentration of surface states. He advocates the thermally stimulated current method to investigate the true density (cm^{-2} eV^{-1}) of electronic states at the interface and in the nitride, and their distribution.

Deal (2) has made extensive dielectric property measurements of silicon nitride films: dielectric constant, breakdown field, surface charge, surface recombination velocity, work function difference, conductance, polarization, and trapping and comparison with silicon oxide films. At the 1966 Electrochemical Society Symposium on Silicon Nitride, it was reported that the silicon nitride-silicon structure gives rise to considerable room temperature instability, other than ion migration, as well as exhibiting a relatively high surface charge (i.e., $> 10^{12}$ cm^{-2}). Experiments conducted by Chu (12) suggest that the silicon-silicon nitride interface is relatively permeable to electrons and/or holes by virtue of either tunnelling or thermal emission processes. Furthermore, the nitride appears to possess sites capable of being charged and discharged, as for example through the filling and emptying of trapping centers. It was not possible to say (in 1967) whether these traps are characteristic of the nitride alone or if the nitride-silicon combination.

An I-V curve is given in Figure III-42 for an electron beam evaporated silicon nitride film (substrate maintained at 250°C).

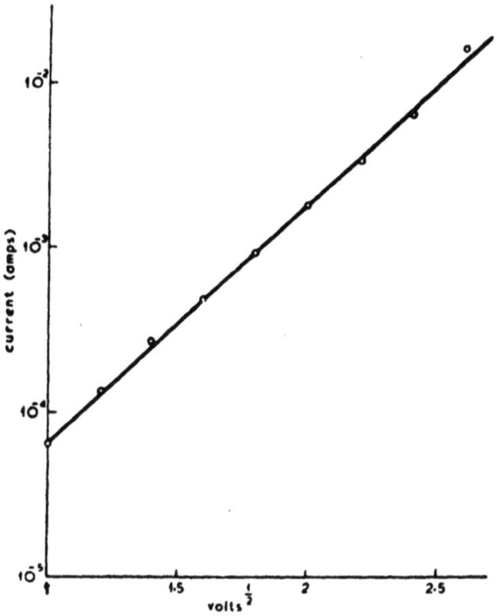

Figure III-42. Typical current-voltage relationship for an evaporated Si_3N_4 film 1650 Å thick. (57)

Deposition pressure $= 10^{-5}$ to 10^{-6} torr Al-Si_3N_4-Al capacitor structure.

Figure III-43 gives dc conduction measurements on sputtered Si_3N_4 films and indicates a linear relation between the logarithm of current and the square root of field.

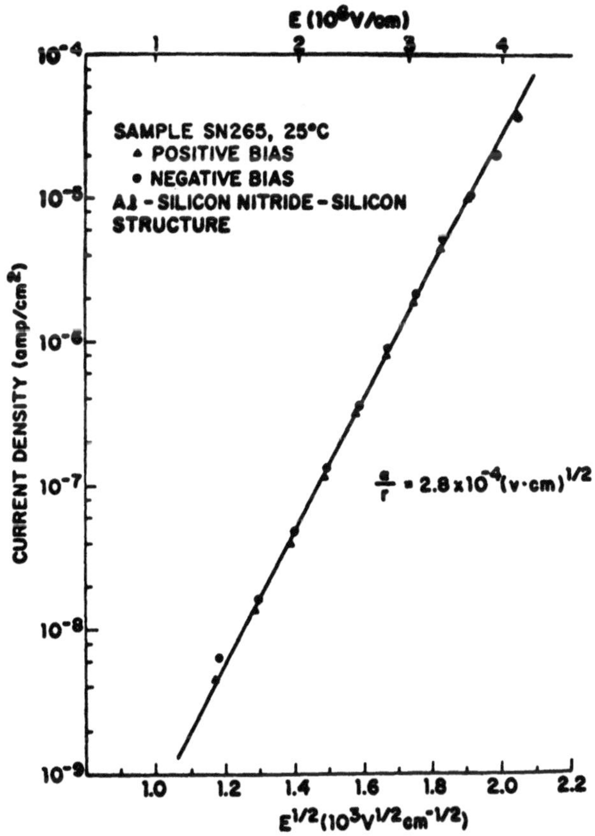

Figure III-43. High-field I-V characteristics of a sputtered silicon nitride film. (65)

The slope of this plot and the independence of current on bias polarity indicated that a bulk-limited conduction mechanism of emission of carriers from traps in the silicon nitride was operative. The dc conduction of the samples was found to be symmetrical with bias and highly nonohmic.

From I-V experiments Sze(59) found the current transport mechanisms in Si_3N_4 films to be bulk controlled. At high temperatures and high fields, the current is mainly due to field-enhanced thermal excitation of trapped electrons into the conduction band. Similar results on this bulk-controlled phenomenon has been observed on reactively sputtered and pyrolytic Si_3N_4 films by Hu (65), and on nitride films formed on Si substrates by silane-ammonia reaction as observed by Brown (61). The latter has commented on the doubtful electrical properties resulting from the deposition of Si_3N_4 directly on Si. Sze's films were deposited by the reaction of $SiCl_4$ with NH_3. He found that at a given temperature and electric field, the current transport depends little on the substrate material, the film thickness, or the polarity of the electrodes. He proposed that the current is bulk controlled and is the sum of three contributions: (a) Poole-Frenkel mechanism, (b) field ionization of trapped electrons, (c) thermal hopping of excited electrons.

Figure III-44 provides data by Doo and Kerr (3) on the I-V (log-log) of a rather thick film (9100 Å) of silicon nitride from which it can be seen that the current changes by seven orders of magnitude for less than a factor-of-three change in bias. The behavior appears to be roughly linear down to extremely low currents.

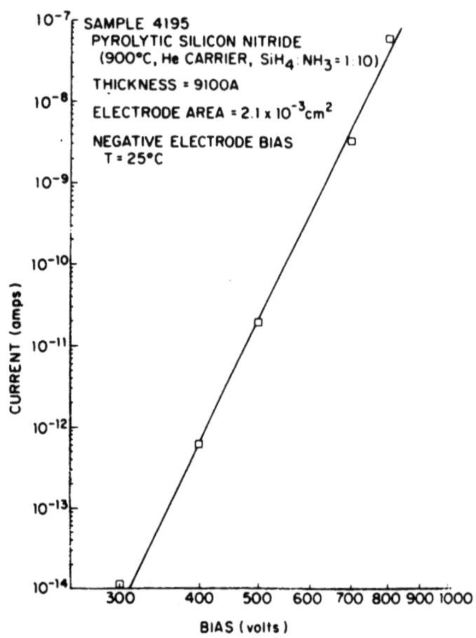

Figure III-44. Typical I-V data for a pyrolytic silicon nitride film. (3)

Even from plots of log I vs $E^{1/2}$ it is apparent that the currents are field controlled rather than voltage controlled for films of varied thickness. When films were prepared by variations in the deposition process (different ratios of SiH_4:NH_3) in H_2 carrier gas at 900°C there was found to exist a shift along the $E^{1/2}$ axis as shown in Figure III-45.

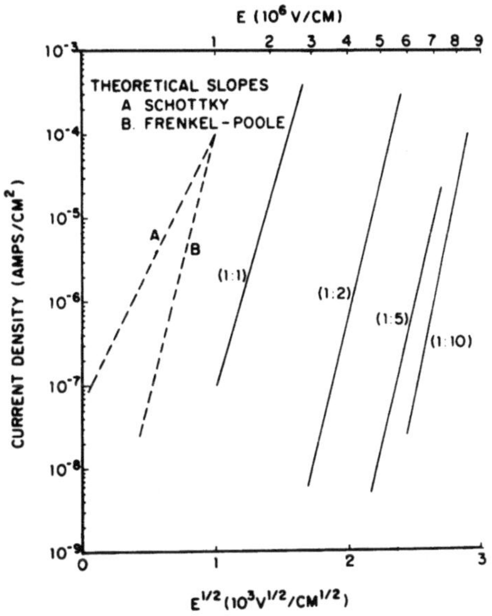

Figure III-45. Electronic leakage in pyrolytic silicon nitride films deposited at 900°C with H_2 carrier gas. Measurements at 25°C, and the silane:ammonia ratios are given in parenthesis on each curve. (3)

C-V Characteristics

The question "Can silicon nitride completely replace silicon dioxide in devices?" has been raised by various investigators. Experiments by Carlson (62) indicate that at least in most applications, this complete replacement will not result in the most desirable device characteristics. His supporting data are shown in Figure III-46.

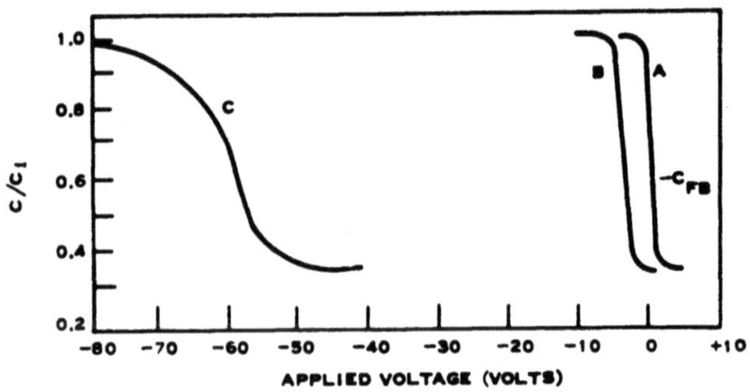

Figure III-46. MIS capacitance-voltage characteristics of structures with equivalent specific capacitance. (62)

CURVE	INSULATOR MATERIALS (THICKNESS)	FLAT BAND VOLTAGE
A	IDEALIZED: $C_I = 2.3 \times 10^{-8}$ fd/cm^2	0
B	SILICON DIOXIDE (1500Å): OR SILICON DIOXIDE (1000 Å) + SILICON NITRIDE (900Å)	−4
C	SILICON NITRIDE (2750Å) ON VAPOR-ETCHED SILICON	−61

SUBSTRATE: P-TYPE SILICON, $N_D = 10^{15}/cm^3$

The high negative flat-band voltage of curve C in Figure III-46 is indicative of a large amount of positive charge associated with the nitride-silicon interface. Such a condition would cause severe inversion of p-type surface regions on transistors or integrated circuits. The smeared-out nature of the C-V relationship is evidence of a high density of chargeable surface states at the nitride-interface. These states can act as traps or recombination centers for carriers being transported in the vicinity of the surface. Such action can result in poor junction characteristics and low device gain. Figure III-47 presents further evidence of electronic-charge instabilities in the nitride-silicon interface.

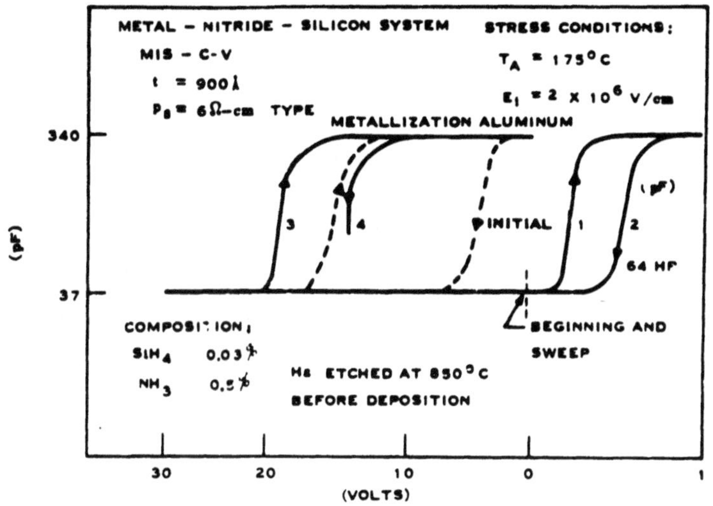

Figure III-47. Metal-nitride-silicon capacitance-voltage and stress characteristics (62)

The large amount of hysteresis and its polarity are suggestive of the electronic-charge injection by tunneling across the nitride-silicon interface and subsequent selective trapping in the nitride (see also the section on I-V Characteristics).

Feist (34) has reviewed the literature on the silane nitridation process for preparing silicon nitride films and noted that the flat-band surface charge was not influenced much, in general, by the NH_3:SiH_4 ratio (ranging from 1:1 to 20:1) when hydrogen or helium was used as a carrier gas (citing Doo's investigations (54). As a function of deposition temperature, the flat-band charge went through a minimum at about 900°C. In contrast, when argon or nitrogen was used as a carrier gas, the flat-band surface charge tended to decrease steadily as the deposition temperature was increased from 700 to 1000°C and, in the case of nitrogen, as the ammonia flow was increased. The smallest flat-band surface charge concentration resulted from the use of nitrogen as carrier gas and was on the order of N_{ss} = 8 x 10^{11}/cm^2 for Si_3N_4 deposited onto bare silicon (SiH_4:NH_3 = 1:20, 1000°C). A thermally grown oxide layer (500 Å thick, grown in situ) placed between the silicon substrate and the nitride layer, reduced the flat-band charge concentration by a factor of 2 to 6.

The deposition of silicon oxynitride on single-crystal silicon (SiH_4:NH_3:O_2 = 1:10:1, nitrogen carrier, substrate temp. = 850°C) led to a surface charge similar to that of pure Si_3N_4. As in the case of pure Si_3N_4, the surface charge was lowered significantly by placing a thermal oxide layer (∿ 150 Å) under the silicon oxynitride.

Silicon oxynitride exhibits a significantly larger threshold field for C-V instabilities at room temperature than silicon nitride (E_{th} = 5.8 - 7.5 x 10^6 V/cm vs E_{th} = 1-2 x 10^6 V/cm) (3). However, when the deposition was carried out onto thermally oxidized silicon, the threshold field was similar in both cases.

In a study of various methods of nitriding silicon halide compounds, Grieco (35) investigated the $SiCl_4/NH_3$ system and found that the flat-band surface state concentration ($N_{ss} \approx$ 5 - 15 x $10^{11}/cm^2$), the breakdown strength ($E_b \approx$ 0.5 - 1 x 10^7 V/cm), and the C-V hysteresis effects exhibited by the resulting silicon nitride film deposited on single-crystal silicon were very similar to those resulting from the silane process.

The effects of a wide range of frequencies on the C-V characteristics at room temperature and at 78°K were studied by Kendall (16) in an attempt to determine the response of the traps (deep and shallow) in terms of the Frenkel-Poole mechanisms of charge transfer. Three types of electronic states were found: interface states (concentration $\sim 10^{10}$ cm^{-2}), shallow nitride traps ($\sim 10^{11}$ cm^{-2}), and deep nitride traps ($\sim 10^{12}$ cm^{-2}). The deep traps were of both types and form compensation was found to occur; thus by applying fields in excess of a certain threshold field ($\simeq$ 1.75 MV cm^{-1}) either a positive or negative charge could be fixed in the nitride. He presented a model for the deep traps as due to stacking faults in local crystallized areas of the amorphous nitride film. A hysteresis effect was observed at all biases and temperatures, its magnitude remaining sensibly constant. This effect he explained in terms of the shallow nitride traps. Figures III-48 and III-49 summarize the high frequency effects on the C-V characteristics.

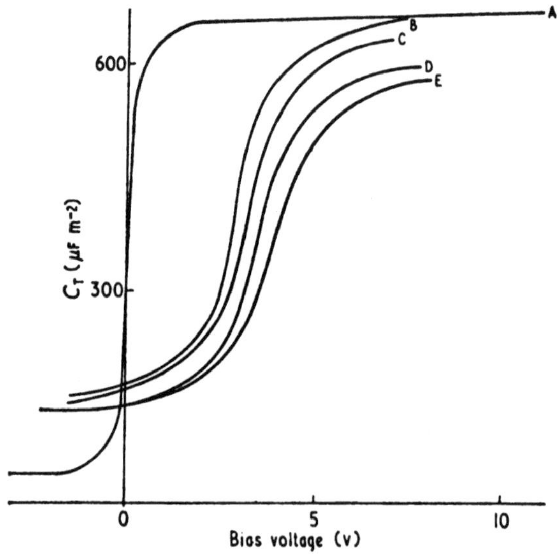

Figure III-48. C-V curves for X11 at room temperature: theoretical curve A based on 10 kHz; experimental curves B 10 kHz, C 100 kHz, D 1 MHz, E 10 MHz. (pyrolytic films) (16)

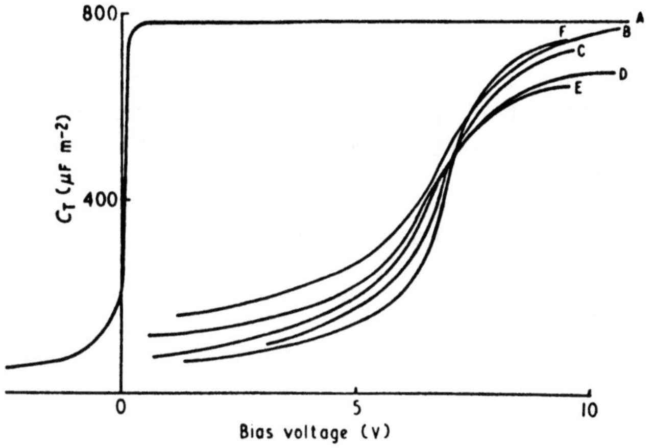

Figure III-49. C-V curves for X11 at 78°K: A, theoretical curve; experimental curves B 500 Hz, C 100 kHz, D 1 MHz, E 500 kHz, F 10 MHz. (pyrolytic films) (16)

Polarization studies by Deal (2) of the silicon nitride layers on n-type Si found the characteristics shown in Figure III-50 to result from stress tests of ± 15 and 30 volts at 300°C for 2 minutes each.

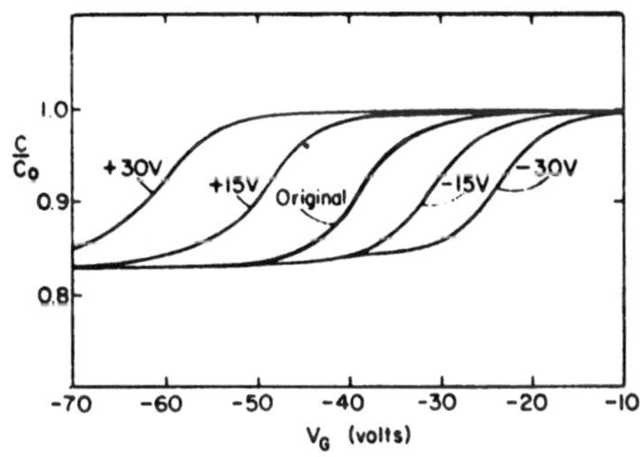

Figure III-50. Example of polarization of silicon nitride layer on silicon due to stress test of ± 15 and 30 v at 300°C for 2 min each. [(111) oriented 1.4×10^{16} cm^{-3} n-type silicon; 0.40 μ nitride thickness.] (2)

The dipolar polarization of the film is self-evident. To study trapping effects, p-type silicon was used with a 0.40-micron silicon nitride thick layer.

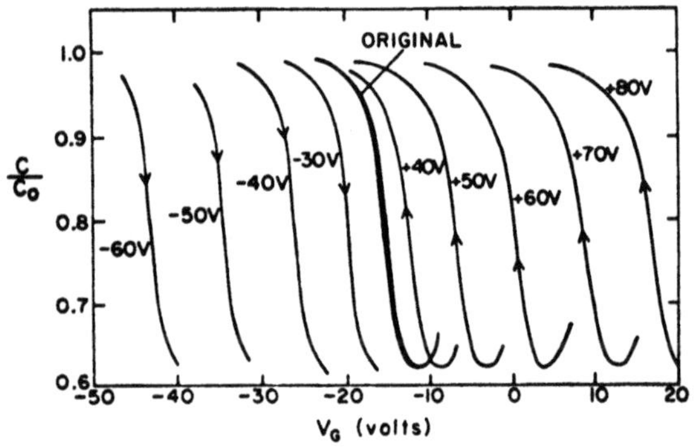

Figure III-51. Capacitance-voltage plots showing example of room-temperature trapping in silicon nitride layer on silicon. [(100) oriented $1.4 \times 10^{16}\ cm^{-3}$ p-type silicon; 0.21 μ nitride thickness.] (2)

Figure III-51 shows that trapping begins at + 40 or -30 volts. For every additional 10 volts applied above the threshold, the C-V characteristic is displaced by about 10 additional volts. A positive applied bias shifts the flatband voltage in the positive direction, demonstrating that the effect is indeed due to trapping and is not caused by ion migration or dipolar polarization.

C-V measurements shown in Figure III-52 on sputtered Si_3N_4 film (MIS) structures reveal hysteresis effects which indicated that carriers injected were predominantly holes rather than electrons. Shifts in the C-V characteristics after bias-temperature stress at 300°C support this finding. Szedon (63) in studying the C-V characteristics of pyrolytic Si_3N_4 films found that both holes and electrons were injected in the MIS structure. The retention of the same sense of the hysteresis curves by increasing the negative bias in going from curve I to curve II indicates holes are being injected. Then, the application of a large positive bias causes a reversal of the sense of the hysteresis in curve III, suggests that hole injection from the metal electrode is predominant in this case.

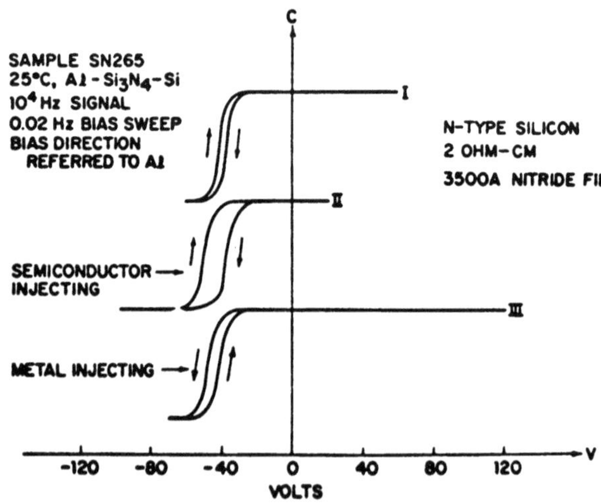

Figure III-52. Experimental C-V traces on Al-Si_3N_4-Si structure with various bias-sweep limits. (11)

C-V data are shown in Figure III-53 for reactively sputtered silicon nitride films for variously treated polished Si surfaces as a means to establish the optimum conditions for the fabrication of a stable nitride structure with a minimum number of surface states (51).

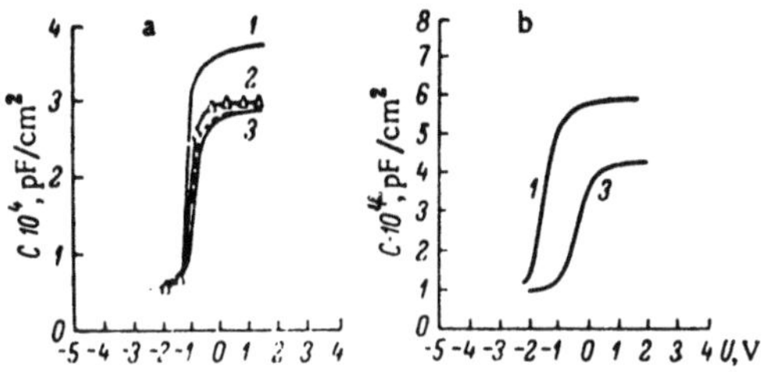

Figure III-53. Capacitance-voltage characteristics C(U) at T = 293°K of metal-dielectric-semiconductor structures prepared from n-type silicon (ρ = 5 Ω . cm) and subjected to various surface treatments: a) mechanically polished surface of silicon without an amorphous film (the dashed curve shows the absence of slow relaxation of the capacitance under the action of a static negative bias); b) mechanically polished surface of silicon with an amorphous film 50 Å thick; f (kc): 1) 2; 2) 465; 3) 1000 (51).

Analysis of the data shows a definite dependence of the capacitance on the frequency and these characteristics also shows some dispersion of the surface of a region depleted of the majority carriers in the frequency range 2 kHz - 1 MHz. Thus, the conditions of preparation of the Si_3N_4-Si interface have a strong influence on the frequency dispersion of the capacitance of metal-dielectric-silicon structures. Further, the annealing of the substrate before growing a film of Si_3N_4 in a vacuum of 10^{-6} torr with a residual nitrogen atmosphere at 500-600°C and subsequent annealing of the Si_3N_4-Si structure under the same conditions reduced the dispersion of the capacitance by stabilizing the interface boundary and the dielectric. The bombardment of the substrate with nitrogen ions before the deposition of a film of silicon nitride and the subsequent annealing of the resultant structure under the foregoing conditions also produced metal-dielectric-semiconductor structures with a small dispersion of the surface capacitance. Finally, measurements of the C-V characteristics carried out after eight months showed that the Si_3N_4-Si interface was stable.

High conduction phenomenon in nitride films have been observed by Doo & Nichols (64) in films obtained from SiH_4/NH_3 and by Hu & Gregor (28). Grieco (35) also experienced high conduction in films prepared by $SiCl_4$ /NH_3 reaction. Figures III-54, III-55 and III-56 summarize his MIS capacitor measurements.

Figure III-54. MIS capacitance curve of a 0.010 in. diameter gold dot on a double dielectric layer (1850 Å nitride on 1000 Å oxide). The nitride was deposited at 1000°C and the silicon was 1 ohm-cm, p-type. The dielectric constant of the composite layer measures 5.0 and the SC + SS density is $3 \times 10^{11}/cm^2$. (35)

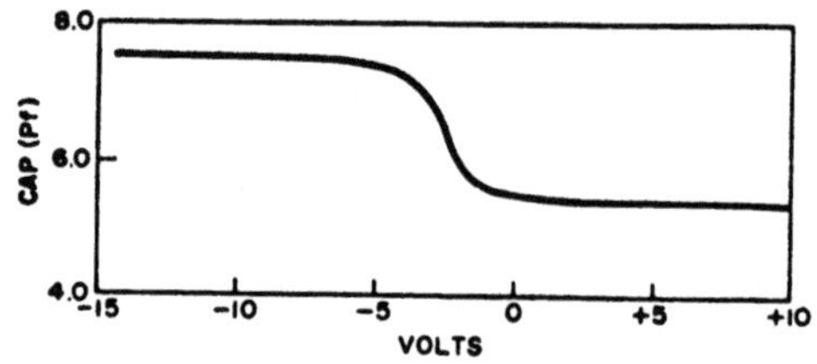

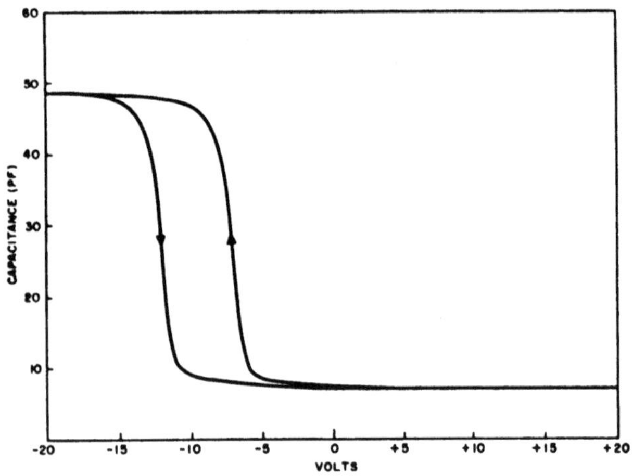

Figure III-55. MIS capacitance curve of a 0.015 in. diameter gold dot on a 1500 Å thick nitride layer deposited on 35-40 ohm-cm, p-type silicon at 1000°C. The dielectric constant is 7.3 and the SC + SS density is $17 \times 10^{11}/cm^2$. (35)

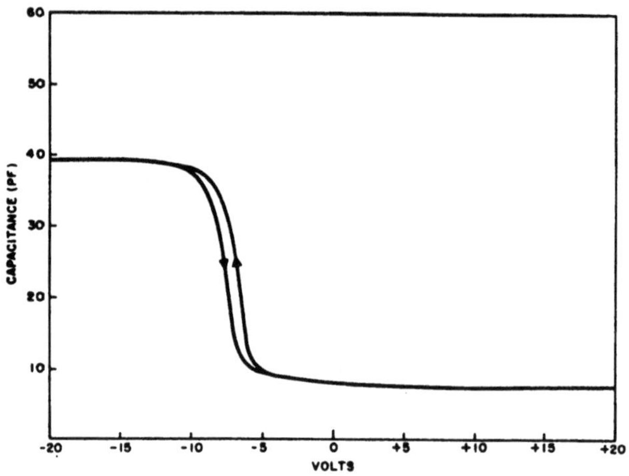

Figure III-56. MIS capacitance curve of a 0.015 in. diameter gold dot on a 1750 Å thick nitride deposited on 35-40 ohm-cm, p-type silicon at 850°C. The dielectric constant is 7.1 and the SC + SS density is $16 \times 10^{11}/cm^2$. (35)

In the latter two Figures, there are indications of an instability in the passivation capabilities of the nitride as evidenced by the hysteresis loop in the C-V plot. These authors believe carrier injection from the silicon, with subsequent trapping in the nitride, is probably one of the reasons for this behavior.

Charge Transport Mechanisms

Charge-transfer mechanisms through silicon nitride films were studied by Kendall (16). He notes that the apparent exponential dependence of the charge transfer on the electric field and the temperature dependence of the mechanism suggests either the Schottky effect or the Frenkel-Poole effect, where the traps emit their carriers to the appropriate energy band of the nitride. However, he states, there are two types of traps, the shallow ones involved in the hysteresis and the deep ones involved in the C-V shift along the bias axis. Above 20 volts, the current will be flowing through all the available traps and below 20 volts it will flow only through the shallow traps. So far, current voltage results indicate a Frenkel-Poole mechanism for charge transfer to and from both types of trap. Current-temperature measurements also indicate some field ionization and trap hopping. Trap-hopping activation measurements by various investigators (Kendall (53), Yeargan & Taylor (6), and Sze (59)) indicate a value about 0.10 eV. Trap levels in the energy band diagram are discussed in the Energy Gap section of this chapter.

Thermally stimulated current (TSC) measurements on silicon nitride* diodes by Kendall (53) have indicated that a complex energetic and spatial trap distribution exists in the films. The basic idea of this experimental procedure is to fill the electronic states under investigation and then cool the system down. The voltage used to fill the traps is then removed and the system warmed up under closely controlled conditions, as the temperature rises the states release their carriers when there is sufficient thermal energy available. A current spectrum is obtained as shown in Figures III-57 and III-58.

* These MIS devices were made using 10 ohm-cm n-type Si which had a Si_3N_4 film deposited on its (111) face. The nitride was deposited either by reacting NH_3 and SiH_4 in a 1:5 ratio by volume at about 850°C, or by using a 1:3 ratio and substrate temperature of 350 to 400°C in rf glow discharge conditions.

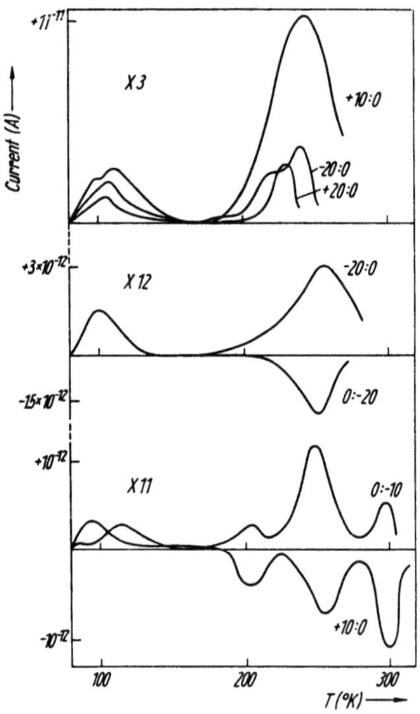

Figure III-57. Typical TSC curves for various biases; β = 0.6°K/sec. Biases; e.g. - 20:0 means - 20 V cooling and 0V warming. (14)

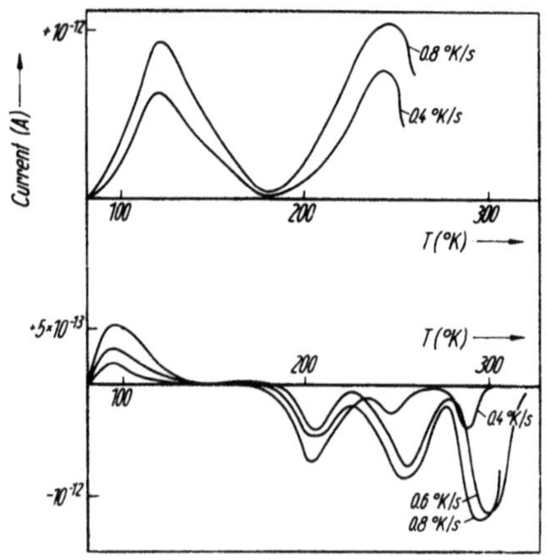

Figure III-58. TSC curves for two samples and for various heating rates, biases: + 10 V cooling and zero-voltage warming. The heating rates are shown in s^{-1} °K/sec for each curve. (14)

Kendall (14) noted that there are certain patterns discernable in Figures III-57 and III-58. For positive warming biases both the broad first peak and the narrow second peak appear, whereas for negative biases only the second peak appears and with a negative current. For zero warming bias, the device behaved as for positive biases until after they had been subjected to high electric fields. His interpretations for this behavior is the following:

> "Since the first peak only occurs when a positive voltage appears across the device, there is either a flow of holes from the silicon nitride and the interface into the silicon, or a flow of electrons from the silicon into the interface and into the nitride. The second peak is due to electrons flowing from the traps in the nitride into either the silicon or the aluminum, depending upon the sign of the applied voltage."

Interesting data on the conduction processes for silicon nitride films deposited on degenerate silicon substrates (0.0005 ohm-cm, p-type) utilizing the pyrolysis reaction of $SiCl_4$ and NH_3 at 1000°C was presented by Sze (59). His plot of the current versus 1/T for a field of 5.3 MV/cm is shown in Figure III-59. The conduction current (J) in the nitride film can be separated into three components: J_1, J_2, and J_3. The current J_1 is due to the Frenkel-Poole emission which dominates at high temperatures and high fields. The current J_2 is due to tunnel emission of x trapped electrons into the conduction band, which dominates at low temperatures and high fields. The current J_3 is the ohmic component which contributes at low fields and moderate temperatures according to this investigator.

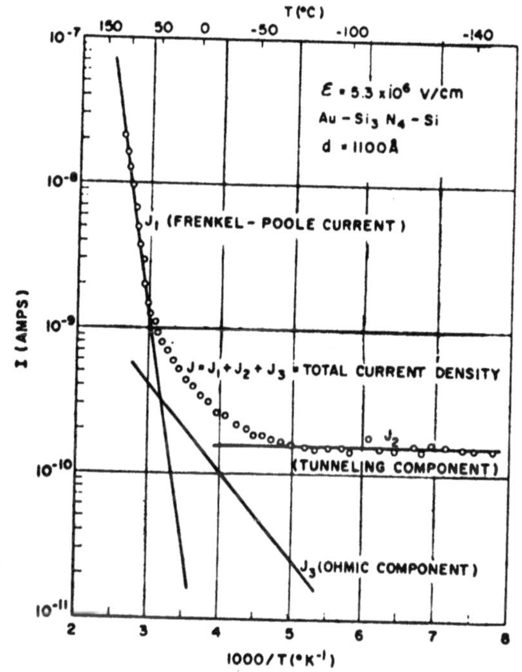

Figure III-59. Current versus 1/T of an Au-Si_3N_4-Si diode at a field of 5.3 x 10^6 V/cm. The total current density is separated into three components: J_1 (Frenkel-Poole current), J_2 (tunnel current), and J_3 (ohmic conduction current). (59)

Sources and Availability of Silicon Nitride

The following companies are suppliers of various forms of silicon nitride:

1. Cerac Inc., Box 126, Butler, Wisconsin 53051 (150 mesh powder)
2. Atomergic Chemetals Co. (Div. Gallard-Schlesinger Chem. Mfg. Co.), 1001 Frankin Avenue, Garden City, L.I., New York. (Si_3N_4 crucibles, mesh powder, lumps)
3. Union Carbide Corp., Carbon Products Div., 270 Park Avenue, New York, N.Y. 10017. (Si_3N_4 crucibles and thermocouple tubes)
4. Union Carbide Corp., Stellite Div., Kokomo, Indiana. (Si_3N_4 refractory parts)
5. Research Organic/Inorganic Chemical Corp., 11686 Sheldon Street, Sun Valley, Calif. 91352.

PROPERTIES-REFERENCES

1. PARR, N.L. Preparation, Micro-Structure and Mechanical Properties of Silicon Nitride. In: SPECIAL CERAMICS, edited by P. Popper. Acad. Press, Inc., New York, N.Y., 1960. p. 102-135.

2. DEAL, B.E. et al. Electrical Properties of Vapor-Deposited Silicon Nitride and Silicon Oxide Films on Silicon. ELECTROCHEM. SOC., J., v. 115, no. 3, Mar. 1968. p. 300-307.

3. DOO, V.Y. and D.R. KERR. Investigation of Refractory Dielectrics for Integrated Circuits. International Business Machines Corp., Cambridge, Mass. NASA-CR-995. Contract No. NAS 12-105. Mar. 1968. Available from Clearinghouse for Federal Scientific and Technical Information, Springfield, Va. 22151.

4. CORDES, L.F. Evidence of Excess Silicon in Reactively Sputtered Silicon Nitride Films. APPLIED PHYS. LETTERS, v. 11, no. 12, Dec. 15, 1967. p. 383-385.

5. SWANN, R.C.G. et al. The Preparation and Properties of Thin Film Silicon Nitrogen Compounds Produced by a Radio Frequency Glow Discharge Reaction. ELECTROCHEM. SOC., J., v. 114, no. 7, July 1967. p. 713-717.

6. YEARGAN, J.R. and H.L. TAYLOR. Conduction Properties of Pyrolytic Silicon Nitride Films. ELECTROCHEM. SOC., J., v. 115, no. 3, Mar. 1968. p. 273-276.

7. COLEMAN, M.V. and D.J.D. THOMAS. The Structure of Amorphous Silicon Nitride Films. PHYS. STATUS SOLIDI, v. 25, 1968. p. 241-252.

8. GYULAI, J. et al. Analysis of Silicon Nitride Layers on Silicon by Backscattering and Channeling Effect Measurements. APPLIED PHYS. LETTERS, v. 16, no. 6, Mar. 15, 1970. p. 232-234.

9. FORGENG, W.D. and B.F. DECKER. Nitrides of Silicon AIME METALLURGICAL SOC., TRANS., v. 242, June 1958. p. 343-348.

10. TURKDOGAN, E.T. et al. Silicon Nitrides: Some Physicochemical Properties. J. OF APPLIED CHEM., (London), v. 8, 1958. p. 296-302.

11. HU, S.M. Properties of Amorphous Silicon Nitride Films. ELECTROCHEM. SOC., J., v. 113, no. 7, July 1966. p. 693-698.

12. CHU, T.L. et al. Preparation and C-V Characteristics of Si-Si_3N_4 and Si-SiO_2-Si_3N_4 structures. SOLID STATE ELECTRONICS, v. 10, Sept. 1967. p. 897-905.

13. BEAN, K.E. et al. Some Properties of Vapor Deposited Silicon Nitride Films Using the SiH_4-NH_3-H_2 system. ELECTROCHEM. SOC., J., v. 114, no. 7, 1967. p. 733-737.

14. KENDALL, E.J.M. Trapping Levels in the Silicon-Silicon Nitride System. PHYS. STATUS SOLIDI, v. 32, 1969. p. 763-768.

15. DALTON, J.V. and J. DROBEK. Structure and Sodium Migration in Silicon Nitride Films. ELECTROCHEM. SOC., J., v. 115, no. 8, Aug. 1969. p. 865-868.

16. KENDALL, E.J.M. The Stabilization of Silicon Surfaces Using Silicon Nitride. BRITISH J. OF APPLIED PHYS., (J. OF PHYS., D), Ser. 2, v. 1, 1968. p. 1409-1420.

17. GREGOR, L.V. Study of Silicon Nitride as a Dielectric Material for Microelectronic Applications. IBM CORP., Interim Report AF33(615)-5386-1. Nov. 15, 1966. 36 p.

18. HARDIE, D. and K.H. JACK. Crystal Structures of Silicon Nitride. NATURE, v. 180, 1957. p. 332-333.

19. SHAFFER, P.T.B. Plenum Press Handbook of High Temperature Materials. V. 1, no. 1, Materials Index. Plenum Press, New York, 1964. p. 283-285.

20. NEWMAN, R. et al. Silicon Nitride for Microelectronics. SPERRY ENGINEERING REVIEW, v. 19, no. 2, 1966. p. 2-9.

21. VAN GELDER, W. and V.E. HAUSER. The Etching of Silicon Nitride in Phosphoric Acid with Silicon Dioxide as a Mask. ELECTROCHEM. SOC., J., v. 114, no. 8, Aug. 1967. p. 869-872.

22. CHU, T.L. et al. The Preparation and Properties of Amorphous Silicon Nitride Films. ELECTROCHEM. SOC., J., v. 114, no. 7, July 1967. p. 717-722.

23. SCOTT, J.H. and J. OLMSTEAD. Meeting Electrochemical Society, Philadelphia, 1966. Extended Abstracts. p. 151.

24. YOSHIOKA, S. and S. TAKAYANAGI. Deposition of Silicon Nitride Films by the Silane-Hydrazine Process. ELECTROCHEM. SOC., J., v. 114, no. 9, Sept. 1967. p. 962-964.

25. KOHLER, W.A. Structural Properties of Vapor Deposited Silicon Nitride. AIME METALLURGICAL TRANS., SOC., v. 246, Mar. 1970. p. 735-740.

26. STERLING, H.F. and R.C.G. SWANN. Chemical Vapour Deposition Promoted by r.f. Discharge. SOLID STATE ELECTRONICS, v. 8, no. 8, Aug. 1965. p. 653-654.

27. JOYCE, R.J. et al. Silicon Oxide and Nitride Films Deposited by an r.f. Glow Discharge. THIN SOLID FILMS, v. 1, no. 6, Mar. 1968. p. 481-494.

28. HU, S.M. and L.V. GREGOR. Silicon Nitride Films by Reactive Sputtering. ELECTROCHEM. SOC., J., v. 114, no. 8, Aug. 1967. p. 826-833.

29. JANUS, A.R. and G.A. SHIRN. Prepartion and Properties of Reactively Sputtered Silicon Nitride. J. OF VACUUM SCI. AND TECHNOL., v. 4, no. 1, Jan./Feb. 1967. p. 37-40.

30. Transetch-N-Selective Etchant for Silicon Nitride Films. TRANSENE COMPANY, INC., Newburyport Turnpike, Route 1, Rowley, Mass. Bulletin No. 122, Jan. 1970.

31. TRIPP, T.B. The Anodic Oxidation of Silicon Nitride Films on Silicon. ELECTRO-CHEM. SOC., J., v. 117, no. 2, Feb. 1970. p. 157-159.

32. SCHMIDT, P.F. and D.R. WONSIDLER. Conversion of Silicon Nitride Films to Anodic SiO_2. ELECTROCHEM. SOC., J., v. 114, no. 6, June 1967. p. 603-605.

33. BROWN, D.M. et al. A New Masking Technique for Semiconductor Processing. ELECTROCHEM. SOC., J., v. 114, no. 7, July 1967. p. 730-733.

34. FEIST, W.M. et al. The Preparation of Films by Chemical Vapor Deposition. In: PHYSICS OF THIN FILMS, edited by G. Hass and R.E. Thun. Acad. Press, New York. 1969. p. 237-322.

35. GRIECO, M.J. et al. Silicon Nitride Thin Films from $SiCl_4$ plus NH_3-Preparation and Properties. ELECTROCHEM. SOC., J., v. 115, no. 5, May 1968. p. 525-531.

36. CERVENAK, J. et al. Some Physical Properties of Si-Si_3N_4 Interfaces and Silicon Nitride Thin Films Prepared by Reactive Sputtering in Nitrogen. J. OF VACUUM SCI. AND TECHNOLOGY, v. 6, no. 6, Nov./Dec. 1969. p. 938-940.

37. OKADA, K. et al. Properties of Thermally Grown Silicon Nitride Films. PHYS. SOC. OF JAPAN, J., v. 23, no. 3, Sept. 1967. p. 655.

38. GERETH, R. et al. New Applications for Low-Temperature Ammonia-Free Si_3N_4 Films. Abstract No. 114. ELECTROCHEMICAL SOCIETY MEETING, Los Angeles, May 10-15, 1970. p. 293-295.

39. REIZMAN, F. and W. VAN GELDER. Optical Thickness Measurement of SiO_2-Si_3N_4 Films on Silicon. SOLID STATE ELECTRONICS, v. 10, no. 7, July 1967. p. 625-632.

40. DRUM, C.M. and M.J. RAND. Chemical Vapor Deposition of a Low-Stress Insulating Film on Silicon. Paper presented at 1968 Electrochemical Society Meeting. ELECTROCHEM. SOC., J., v. 115, 1968. p. 194C.

41. DRUM, C.M. and M.J. RAND. Recent News Item, Item 313, 1968 Electrochemical Society Meeting, Boston.

42. TOKUYAMA, T. et al. Thermal Expansion Coefficient of a Pyrolytically Deposited Silicon Nitride film. JAPAN. J. OF APPL. PHYS., v. 6, no. 10, 1967. p. 1252-1253.

43. DOO, V.Y. et al. Preparation and Properties of Pyrolytic Silicon Nitride. ELECTROCHEM. SOC., J., v. 113, no. 12, Dec. 1966. p. 1279-1281.

44. PHILIPP, H.R. Electrochemical Society Meeting, Philadelphia. Extended Abstracts 155, 1966.

45. BURKHARDT, P. and R.F. MARVEL. Thermal Expansion of Sputtered Silicon Nitride Films. ELECTROCHEM. SOC., J., v. 116, no. 6, June 1969. p. 864-866.

46. POPPER, P. and S.N. RUDDLESDEN. The Preparation, Properties and Structure of Silicon Nitride. BRITISH CERAM. SOC., TRANS., Sept. 1961. p. 603-624.

47. WELLS, W.M. Silicon Nitride as a High-Temperature Radome Material. University of California, Lawrence Radiation Laboratory, Livermore, Calif., Contract W-7405-eng-48, UCRL-7795, May 19, 1964. 19 p.

48. COLLET, M.G. Depositing Silicon Nitride Layers at Low Temperatures Using a Photochemical Reaction. ELECTROCHEM. SOC., J.,-SOLID STATE SCIENCE, v. 116, no. 1, Jan. 1969. p. 110-111.

49. FELDMAN, C. and M. HACSKAYLO. Temperature Characteristics of Vacuum Deposited Dielectric Films. REV. OF SCIENTIFIC INSTRUMENTS, v. 33, Dec. 1962. p. 1459-1460.

50. BARNES, C.R. and C.R. GEESNER. Silicon Nitride Thin Film Dielectric. ELECTROCHEM. SOC., J., v. 107, no. 2, Feb. 1960. p. 98-100.

51. ALEKSANDROV, L.N. et al. Electrical Properties of an Si_3N_4-Si Interface, Prepared by the Method of Reactive Sputtering of Si in a Nitrogen Atmosphere. SOVIET PHYS. SEMICONDUCTORS, v. 3, no. 10, Apr. 1970. p. 1329-1330.

52. KUWANO, Y. The Preparation and Properties of Silicon Nitride Produced by a Radio Frequency Glow Discharge Reaction of Silane and Nitrogen. JAPAN. J. OF APPL. PHYS., v. 7, no. 1, Jan. 1968. p. 88.

53. KENDALL, E.J.M. The Conduction Processes in Silicon Nitride. CANADIAN J. OF PHYS., v. 46, 1968. p. 2509-2516.

54. DOO, V.Y. et al. Property Changes in Pyrolytic Silicon Nitride with Reactant Composition Changes. ELECTROCHEM. SOC., J., v. 115, no. 1, Jan. 1968. p. 61-64.

55. YAMAZAKI, S. et al. Silicon Nitride by Using Strain Films on Silicon. JAPAN. J. OF APPL. PHYS., v. 6, no. 3, 1967. p. 408.

56. TOMBS, N.C. et al. A New Insulated-Gate Silicon Transistor. IEEE PROC., (LETTERS), v. 54, Jan. 1966. p. 87-89.

57. ELLIOTT, E. et al. Some Properties of Electron Beam Evaporated Si_3N_4 Films. THIN SOLID FILMS, v. 3, 1969. p. R47-R48.

58. SCHAFFER, P.S. and B. SWAROOP. Vapor Phase Growth and dc Breakdown of Silicon Nitride Films. CERAMIC BULLETIN, v. 49, no. 5, 1970. p. 536-538.

59. SZE, S.M. Current Transport and Maximum Dielectric Strength of Silicon Nitride Films. J. OF APPLIED PHYS., v. 38, no. 7, June 1967. p. 2951-2956.

60. YEARGAN, J.R. et al. A Charge Transport Mechanism in Thin Silicon Nitride Films. Univ. of Texas, 1967. 120 p. NASA N67-37944.

61. BROWN, D.M. et al. Properties of $Si_xO_yN_z$ Films on Si. ELECTROCHEM. SOC., J., v. 115, no. 3, Mar. 1968. p. 311-317.

62. CARLSON, H.G. et al. Development of Silicon Nitride and Cermet Resistors For Use in a Binary Counter, Metal Insulator Field Effect Transistor Circuit. Texas Instruments Report No. 03-68-42, Mar. 1968. 147 p. NASA N69-10283.

63. SZEDON, J.R. et al. Charge Instability in Metal-Silicon Nitride-Silicon Structures. IEEE SOLID STATE DEVICE RESEARCH CONFERENCE, June 1966.

64. DOO, V.Y. and D.R. NICHOLS. ELECTROCHEM. SOC., J., v. 113, 1966. p. 212C.

65. HU, S.M. et al. Evidence of Hole Injection and Trapping in Silicon Nitride Films Prepared by Reactive Sputtering. APPLIED PHYS. LETTERS, v. 10, no. 3, 1967. p. 97-99.

SILICON NITRIDE-SILICON DIOXIDE COMBINATIONS

Introduction

While it would be desirable to replace completely SiO_2 by Si_3N_4, such structures exhibit hysteresis in surface behavior and charge instability due to the high number of surface charges, tunneling and trapping at the $Si-Si_3N_4$ interface as noted in the previous sections of this survey. Two major approaches have been taken to solve these problems: (1) mixed oxy-nitride system, (2) nitride-oxide layer composite structure. Each approach will be discussed separately and has been investigated by numerous investigators.

Mixed Oxy-nitride System Approach

Various structures of $SiO_2-Si_3N_4$ mixtures, rather than pure Si_3N_4 have been investigated since they are easier to etch, but they have been found to be inferior to the oxide-nitride sandwich structure (1,2).

Chu and coworkers (1) have studied films of $Si_3N_4-SiO_2$ mixtures on silicon substrates in an effort to obtain a compromise between the deficiencies of each of these commonly used dielectric materials:

SiO_2

1) SiO_2 is structurally porous
2) SiO_2 has high permeability toward impurities and the migration of impurity ions in silica films are undesirable in device applications
3) density = 2.2 g/cm^3 for thermally grown SiO_2
6) Low dielectric constant (approx. 3.9)

Si_3N_4

1) Si_3N_4 is more dense and more impervious than SiO_2
2) Ion-migration effect is negligible even at high temperatures in Si_3N_4 films
3) density = 2.8 to 3.0 g/cm^3 for amorphous Si_3N_4
4) Si_3N_4 is not compatible with Si in properties
5) $Al-Si_3N_4-Si$ (MNS) structures have been shown to possess inherently unstable charge distribution under the application of large insulator fields ($\pm 6 \times 10^6$ V/cm) at room temperature
6) High dielectric constant (approx. 8)

The properties of these films, such as composition, density, dissolution rate, dielectric constant, refractive index, and infrared absorption were studied as a function of the reactant composition. The mixed nitride-oxide films were prepared by the reaction of silane with an ammonia-oxygen mixture and deposited on single crystal silicon substrates (111) orientations (n-type, 20-50 ohm-cm resistivity). Graphs summarizing the density, refractive index, and dissolution rate of the deposit as a function of the O_2/SiH_4 molar ratio in the reactant are presented by Chu. (1)

Brown and coworkers (2) have also studied the physical, electrical and optical properties of $Si_xO_yN_z$ films on silicon which were formed by the pyrolysis of various mixtures of SiH_4, NH_3, and NO. They found that the physical properties of glassy-amorphous oxynitride films on silicon are strongly dependent on the oxygen content.

Doo and Kerr (3) also reported on the growth rate of the silicon oxynitride as a function of temperature and O_2 flow rates for the growth conditions shown in the graph legends of Figures IV-1 and IV-2.

The etch rate of silicon oxynitride in 7:1 buffered HF solution was investigated by Doo & Kerr (3) as a function of substrate temperature and the oxygen flowrate during film growth. Figure IV-3 shows the resulting etch rate vs inverted temperature plot. This curve is quite similar to the etch rate curve for silicon nitride in 48% HF also determined by these authors. The high temperature films etched slower, primarily due to higher density, than the low temperature films. They also reported on the change of the etch rate as a function of oxygen flowrate during film growth as shown in Figure IV-4.

The etch rate increases rapidly with a small increase of oxide in the low oxide region; thereafter, the rate of increase gradually decreases. When the films contain mostly oxide (at 3 ml/min. O_2), the etch rate approaches its peak.

Figure IV-1
Growth rate of silicon oxynitride vs. temperature (3)

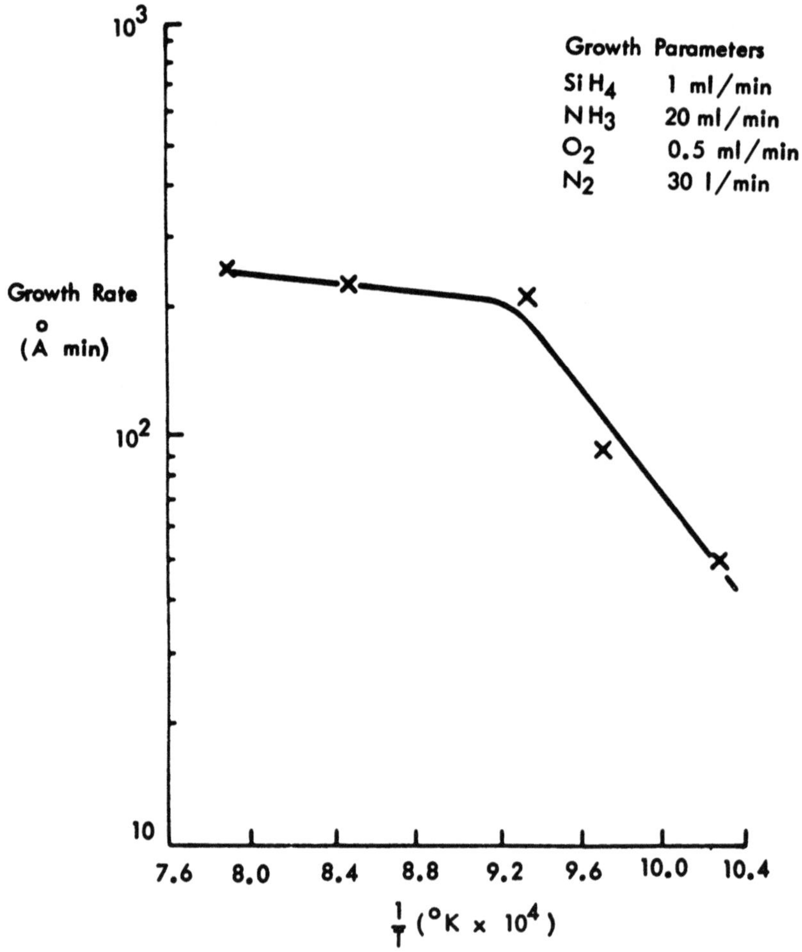

Figure IV-2
Growth rate of silicon oxynitride vs. O_2 flowrate (3)

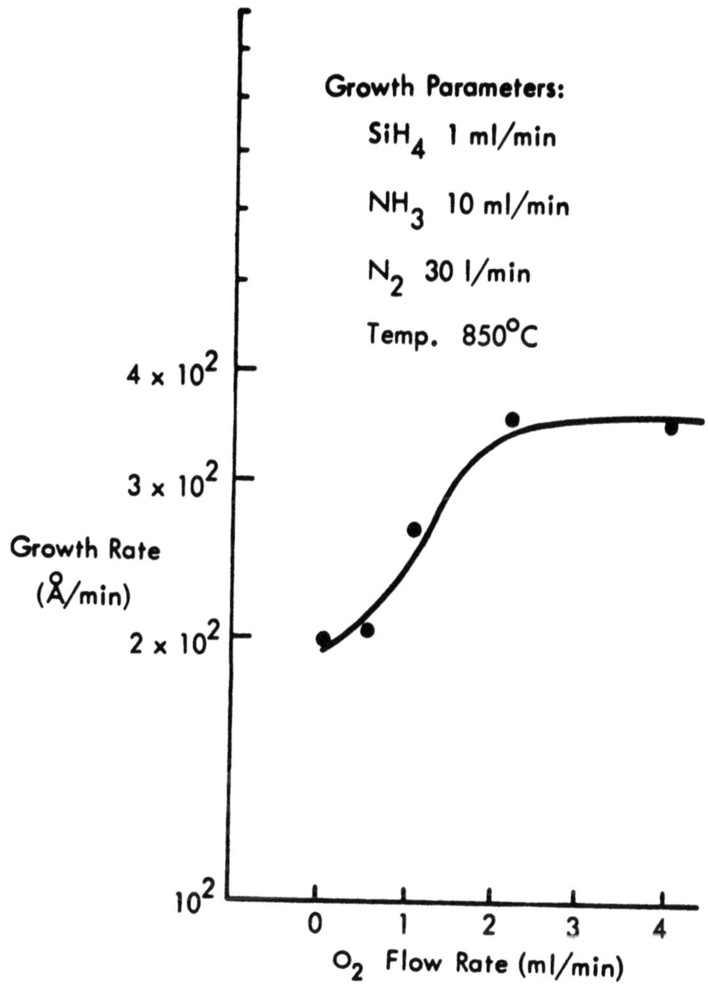

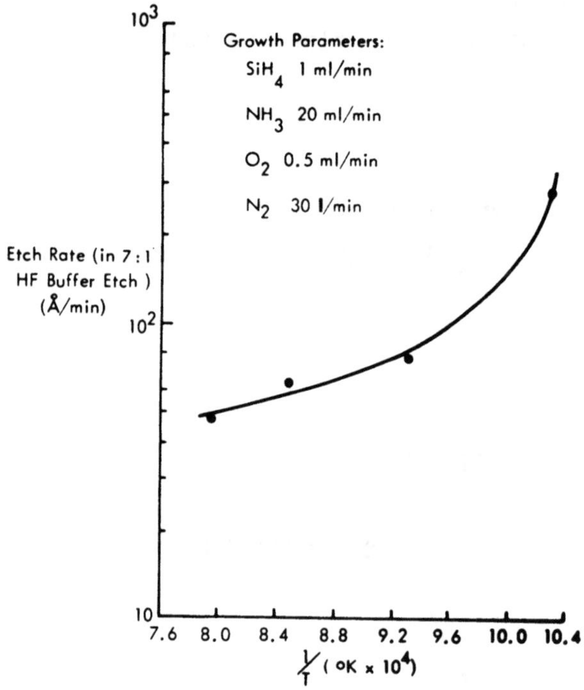

Figure IV-3
Etch rate of silicon oxynitride in 7:1 buffered HF solution (7 parts 42% NH_4F and 1 part 50% HF) vs. temperature (3)

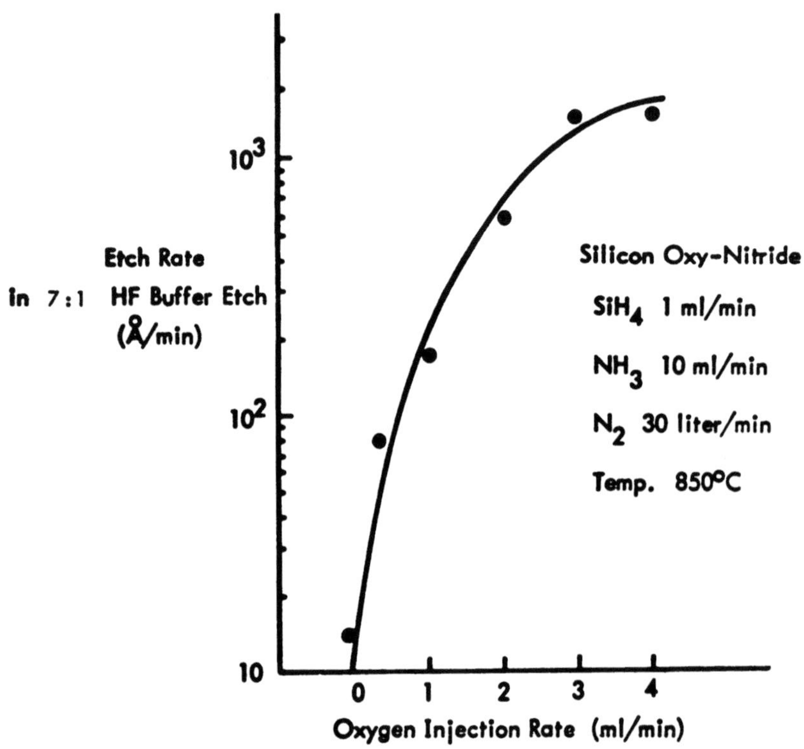

Figure IV-4
Etch rate of silicon oxynitride in 7:1 buffered HF solution (7 parts 42% NH_4F and 1 part 50% HF) vs. O_2 flowrate (3)

Refractive index data on silicon oxynitride films reported by Doo & Kerr (3) as a function of growth temperature and oxygen flowrate are given in Figures IV-5 and IV-6.

Figure IV-5
Refractive index of silicon oxynitride vs. temperature (3)

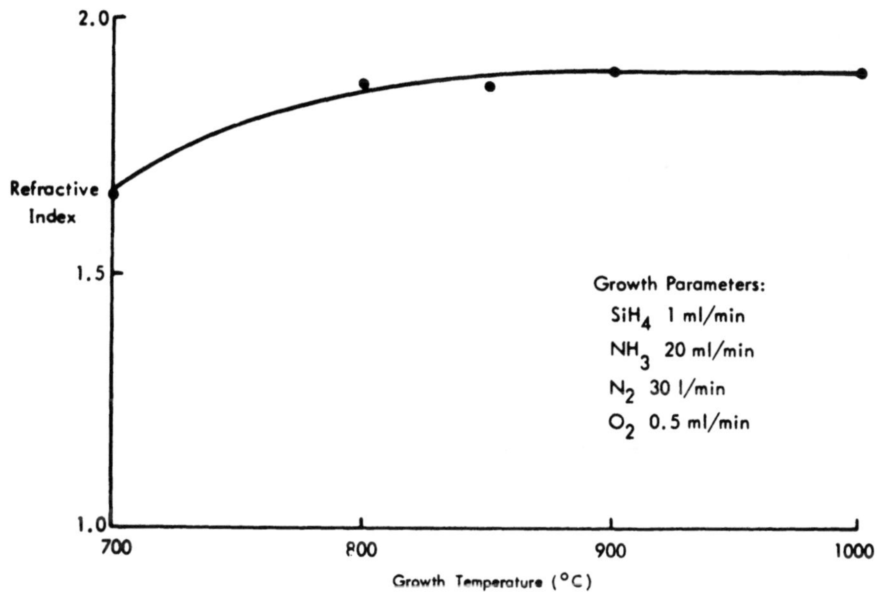

Figure IV-6
Refractive index of silicon oxynitride O_2 flowrate (3)

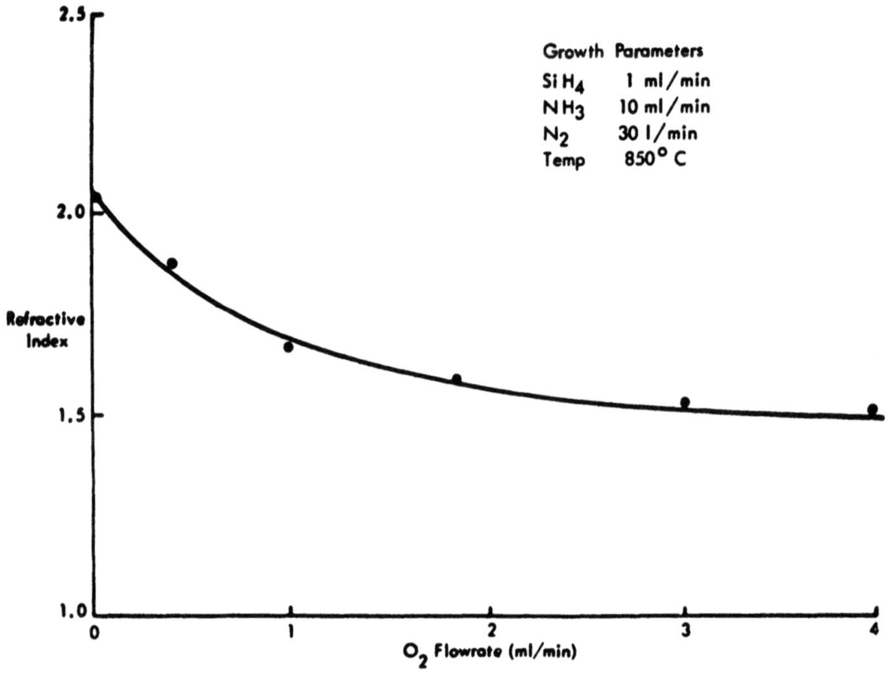

With silicon dioxide having a refractive index of approx. 1.45 and silicon nitride approx. 2.00 the above data are in agreement----i.e., lower temperatures tending to more oxide formation. Further, as the flowrate of oxygen increases while that of silane and ammonia are constant, the oxide content would obviously increase.

Doo & Kerr (3) have conducted a limited number of electrical measurements on a number of oxy-nitride films to determine the effect of film deposition temperature and the effect of an underlying thermal oxide. Table IV-1 summarizes the limited data available.

Table IV-1. Electrical Measurements on Nitride-Oxide Mixtures (3)

Wafer No.	4389	4390	4393	4394	4395	4396
Thermal oxide	none	150A	none	none	none	none
Nitride-Oxide Deposition Temp(°C)	850	850	700	800	900	1000
$SiH_4:NH_3:O_2$	1:10:1	1:20:1	———	1:20:0.5	———	
Dielectric constant	-	-	4.8	5.5	5.2	6.1
Rm-temp. drift threshold (10^6V/cm)	6-7	5	3.1	1.6	4.5	4.0
$-N_{FB}$ before anneal ($10^{11}/cm^2$)	24	6.5	16	19	13	18
$-N_{FB}$ after 300°C, N_2, 30m ($10^{11}/cm^2$)	18	5.7				
ΔN_{FB}, + bias,* 200°C, 30 m ($10^{11}/cm^2$)	-0.3	-0.2				
ΔN_{FB}, + bias,* 200°C, 20 h ($10^{11}/cm^2$)		-0.6				
ΔN_{FB}, - bias,* 200°C, 30 m ($10^{11}/cm^2$)	-1.8	-0.6				
ΔN_{FB}, - bias,* 200°C, 20 h ($10^{11}/cm^2$)		-1.1				

*Bias gives D = $\pm 3.4 \times 10^{-7}$ Coul/cm^2.

The following conclusions resulted: The dielectric constants for the mixtures are between those of SiO_2 and Si_3N_4, as expected. Room-temperature drift thresholds are well above those of pure nitride (1-2 x 10^6 V/cm), except for one anomalous wafer. The Δ N_{FB} shifts are comparable to those observed with pure nitride.

Rand and Roberts (4) described a chemical vapor deposition process capable of depositing films ranging from SiO_2 to $SiON_{0.67}$ at 700-900°C. Film composition is controlled by the ratio of the reactants, and deposition rates of 200-2000 Å/min may be selected. A preferred film, $Si_3O_4N_2$, is readily etched in buffered HF, but is nevertheless a fairly good sodium ion barrier.

Frank and Moberg (5) sputtered silicon oxynitride by rf reactive sputtering of silicon in mixtures of N_2 + O_2 as part of a program to investigate new dielectric materials for use in passive components in integrated circuits, particularly for metal-insulator-metal (MIM) capacitors. Measurements of the electrical and physical properties of the sputtered oxynitride films were carried out as well as with additions of N_2O in N_2 in concentrations ranging from 0 to 20%. The results of these measurements are summarized in Table IV-2 and Figures IV-7, IV-8, and IV-9.

Table IV-2. Effect of N_2O additions on the dielectric properties of silicon oxynitride films (5)

% N_2O	Etch rate, A/min	IR absorption, maximum (μ)	Breakdown field,* V/cm × 10^6	Dielectric constant, at 1 kHz	Figure of merit**
0	20	12.1	6.0	9.1	5.4
1	42	12.0	6.9	8.7	6.0
3	143	11.8	9.3	8.3	7.7
4	167	11.2	10.1	7.2	7.3
5	225	11.0	10.1	6.8	6.9
7	305	10.7	10.1	6.1	6.2
10	587	10.1	10.9	5.3	5.8
20	1180	9.6	8.3	4.0	3.3
(50% O_2, 50% Ar)	1300	9.3	8.0	3.9	3.1

*At the center of higher peak on breakdown distribution curve for MIM capacitor (see Figure IV-9).
**Product of ε' x dielectric strength

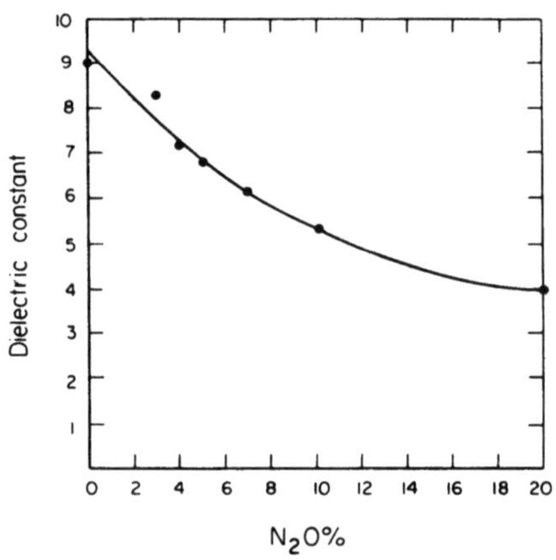

Figure IV-7. Dielectric constant of sputtered silicon oxynitride as a function of percent of N_2O in the sputtering gas. (5)

Figure IV-8

Infrared transmission spectrum of sputtered silicon oxynitride as a function of N_2O% in the sputtering gas. (5)

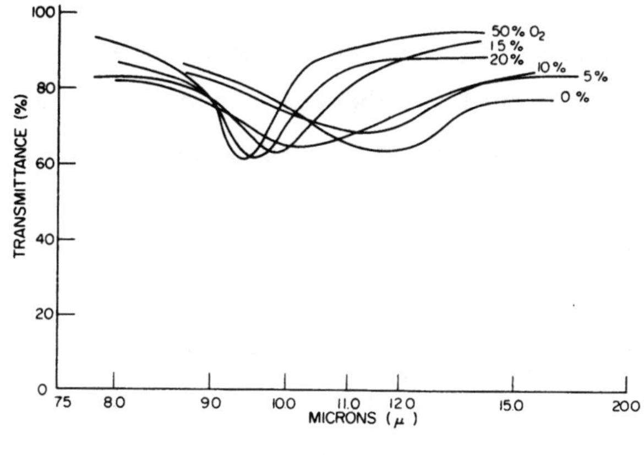

Figure IV-9

Effect of N_2O additions on the breakdown voltage distribution curves for MIM capacitors. (Dielectric thickness 1500 Å, Mo top and bottom electrodes. (5)

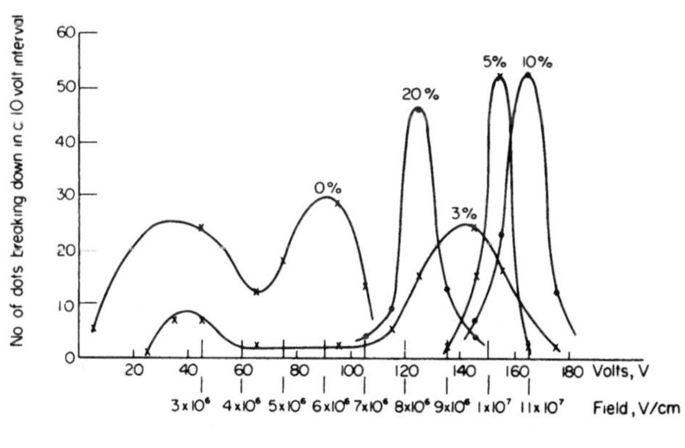

From Table IV-2, it is apparent that the dielectric strength is at a maximum in the range of 5-10% N_2O; similar results apply to the capacitor yield (breakdown strength distribution) which is defined by these authors as the percentage of dots breaking down within a given range of the most probable breakdown (as shown in Figure IV-9).

Frank and Moberg (5) also made high-frequency measurements of capacitance and loss tangent on devices with individual molybdenum top and bottom electrodes, which had a dielectric sputtered in the presence of 5% N_2O added to the nitrogen. The loss tangent of these devices was 0.003 at 1 kHz, 0.0015 at 10 kHz, and 0.01 at 1 MHz. The dielectric constant changed by only 3% in going from 100 Hz to 1 MHz. The temperature coefficient of capacitance is +150 ppm/°C between 25° and 125°C at 1 and 10 kHz. Thus, the measurements show these devices to have highly desirable properties as thin film capacitors. They concluded that rf sputtering of Si in 5% N_2O + 95% N_2 mixtures yields a silicon oxynitride with many desirable properties for use in integrated circuit applications. Principally, these are high breakdown strength, high yield, a dielectric constant of 6.8, and an etch rate of approx. 200 Å/min. Due to the increased breakdown and dielectric constant, the maximum capacitance per unit area is about 2.5 times that of sputtered silicon dioxide. In relation to pure silicon nitride (0% N_2O), the most striking features are the increased etch rate, increased breakdown strength, and high yield as shown by the sharply peaked breakdown distribution curve in Figure IV-9.

Kendall (6) has called attention to the problem of the presence of oxygen in small quantities in silicon nitride films because they lead to the formation of Si_2ON. Forgeng and Decker (7) determined this latter compound to be orthorhombic in structure and is often mistaken for a third (α) phase of Si_3N_4 according to Kendall (6).

Drum and Rand (8) prepared oxy-nitride films on silicon in an effort to produce low-stress films as stresses may cause film cracking, elastic distortion of thin substrates, and plastic deformations in the silicon. They felt that since silicon dioxide and silicon nitride show stresses of opposite sign, there should be an intermediate oxy-nitride composition with nominally zero stress. Their approach was making such a film by the chemical vapor deposition process.

Tombs and coworkers (9) have investigated the preparation and properties of the solid solution series between silicon nitride and silicon dioxide and members. They observed that the pyrolytic reaction between silane and ammonia gases can, in the presence of certain oxygen-containing gases, yield amorphous films having compositions intermediate between those of silicon nitride and silicon dioxide. They note that these materials are to be regarded as solid solutions rather than multiphase mixtures as shown by infrared absorption, electron diffraction, and optical studies.

Nitride-Oxide Composite Layer Sandwich Approach

Jones and Doo (10) have noted that it is also possible to fabricate composite layers of silicon nitride and silicon dioxide. This may be the most advantageous structure since thick layers (greater than 1 μ) of silicon nitride are prone to cracking. For the composite structure, a thin layer of nitride provides the chemical barrier against HF attack or contamination with alkali ions, while a thick layer of oxide lowers the capacitance linkage across the isolation dielectric layer.

Savage (11) found that a layer of SiO_2 between the Si_3N_4 and Si improved the adhesion characteristics of the Si_3N_4. Such a composite layer provides a continuous interface which is adherent at high temperatures, and which remains impervious to diffusion.

Interest in the Si-SiO_2-Si_3N_4 structure has stemmed largely from the charge instability problem and trapping centers associated with the Si-Si_3N_4 interface. Since Si-SiO_2 structures exhibit no charge instability at room temperature, it was felt by Chu (12) that the insertion of a thin SiO_2 layer in the Si-Si_3N_4 structures could minimize the charge instability problems. Additionally SiO_2 has a smaller thermal expansion coefficient than Si, the interface is less stressed than the nitride-silicon interface.

Chu (12) also prepared Si-SiO_2-Si_3N_4 (MNOS) structures using silane + NH_3 + O_2 and evaluated the insulator charge behavior by C-V measurements. The SiO_2 layers ranged in thickness from 70 to 200 Å while the amorphous Si_3N_4 overlayer was fixed at 1000 Å. Bias-stress was applied at room temperature, and the flat-band voltage was subsequently measured to determine the charge state of the insulator. The SiO_2 interlayer was found to be capable of suppressing the charge instability problem due to tunneling and trapping.

Frohman-Bentchkowsky and Lenzlinger (13) made MNOS structures on 200-μ thick Si wafers <100> oriented with a resistivity range of 5-8 Ω cm. Thermal oxide layers of different thickness were grown at 920°C in a dry oxygen ambient. Si_3N_4 was deposited by decomposition of $SiCl_4$ in the presence of NH_3 at 800°C with H_2 as the carrier gas. Different Si_3N_4 current-field characteristics were obtained by variation of the $SiCl_4$-to-NH_3 ratio.

Wang and Van Buren (14) have evaluated Si_3N_4 and Si_3N_4/SiO_2 composite layers as crosspoints in switching matrices. The purpose of this investigation was to fabricate conductor/insulator/conductor structures and to evaluate them as bistable switching elements. No rating of the Si_3N_4 was made against the numerous other structures tested because the current solid state technology was not sufficient to develop practical dielectric thin film devices as bistable switching elements for high frequency, high speed, good isolation telecommunication systems.

Swaroop (15) investigated the dc conduction and breakdown of composite layers of silicon dioxide-nitride (MNOS) between 298°K and 473°K. He found from the current-voltage measurements for various composite structures of oxide-nitride, the conduction is continuous and several orders less than in nitride films alone. However, this current continuity in composites is due to voltage sharing capacity of oxide and nitride, at a given current density. These results support the steady-state current flow model as proposed by Frohman-Bentchkowsky and Lenzlinger (13). The maximum dielectric breakdown strength of composite structures has shown tremendous increase over those of nitride films alone. From HTRE (high temperature reverse biased) tests, the packaged pin devices insulated with composite layers yielded far better stable performance than those devices with oxide alone. Schaffer and Swaroop (16) presented breakdown voltage data on oxide-nitride composite layers of various thicknesses as shown in Table IV-3.

Table IV-3. Breakdown Voltage for Various Thicknesses of Silicon Nitride and Silicon Dioxide-Nitride Films (16)

Specimen No.	Substrate	Thickness (Å) Oxide	Nitride	Deposition temp. (Optical pyrometer) (°C)	Maximum breakdown voltage Si(−)	Si(+)
1	*b*		5000	950	530	574
2	*a*		5000	950	580	658
3	*a*		5000	1000	917	1000
4	*b*		5000	1000	640	680
5	*a*		5000	1100	658	430
6	*b*	1200(*v*)	2500	1000	353	530
7	*a*	1200(*v*)	4000	1000	505	528
8	*b*	1200(*t*)	2500	1000	503	585
9	*a*	1200(*t*)	3000	1000	802	863
10	*a*	1200(*t*)	5300	1000	192	245
11	*a*	1200(*t*)	4000	900	411	523
12	*a*	1200(*t*)	4000	950	545	650
13	*a*	1200(*t*)	4000	1000	970	1020
14	*a*	1200(*t*)	4000	1100	1050	1100

NOTE: *b*—Silicon doped with 10^{15} atoms antimony/cm^3 silicon
a—Silicon doped with 5×10^{18} atoms antimony/cm^3 silicon
v—Vapor-deposited oxide
t—Thermal oxide

From Table IV-3, it appears that the thickness of silicon nitride over oxide affects the breakdown strength of the structure. For example, a 5000 Å thick nitride film over a 1200 Å thick oxide layer gave a very low breakdown voltage (192-245 V at 1 nA). Examination of this thick structure under an optical microscope revealed that the nitride layer had developed microcracks on the surface. This may be the reason, according to these investigators, that breakdown strength for 5000 Å thick nitride films were observed to be lower than that of thinner films.

Deal (17) has investigated C-V plots for double layers of silicon nitride over thermal oxide and compared them with the individual materials in order to elucidate the surface charge effects in the silicon nitride. Measurements were made at 100 kHz and all structures were prepared using (111), 1.4×10^{16} cm^{-3} p-type Si substrates and circular (375 micron dia.) Al field plates. After metalization, they were annealed at 550°C for 2 minutes in dry N_2. From the data, it is evident that the nitride film results in a relatively high surface charge, $Q_s/q > 10^{12}$ cm^{-2}, compared to 6×10^{11} cm^{-2} for the vapor-deposited oxide and 2×10^{11} cm^{-2} for the thermal oxide. The nitride-over-oxide structure has a charge just slightly higher than that of the plain thermal oxide as shown in Figure IV-10.

Figure IV-10. Typical capacitance-voltage plots for MIS structures (17)

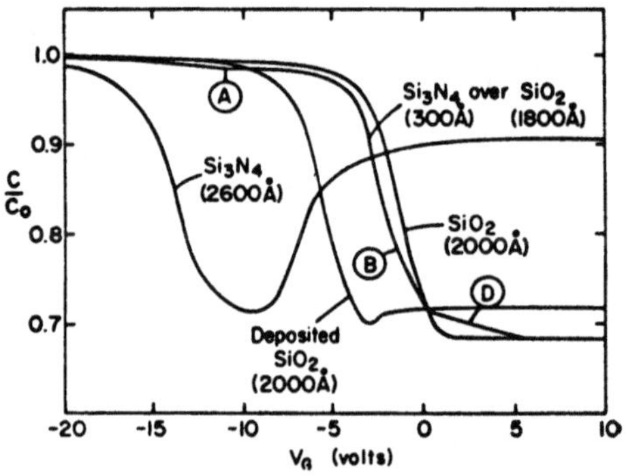

It has been found that if a 0.1-micron thermal silicon oxide layer is present between the silicon nitride and the silicon, no trapping is observed even for voltages up to destructive breakdown of the dielectric layer. However, if the thermal oxide is thin, i.e., 50 to 500 Å, trapping can still occur even though higher threshold fields are required. This increase in threshold field is proportional to the thermal oxide thickness and is considerably larger for negative bias than for positive bias. Data from Deal (17) shows that as the oxide thickness increases from 0 to 500 Å, the positive threshold voltage for a 0.25-micron silicon nitride film increases from 40 to 130 volts. Figure IV-11 shows that the magnitude of the negative threshold voltage increased from 30 to 230 volts as the oxide thickness is increased from 0 to 290 Å. The figure also shows that a sample with a 500 Å oxide breaks down before negative bias trapping is observed.

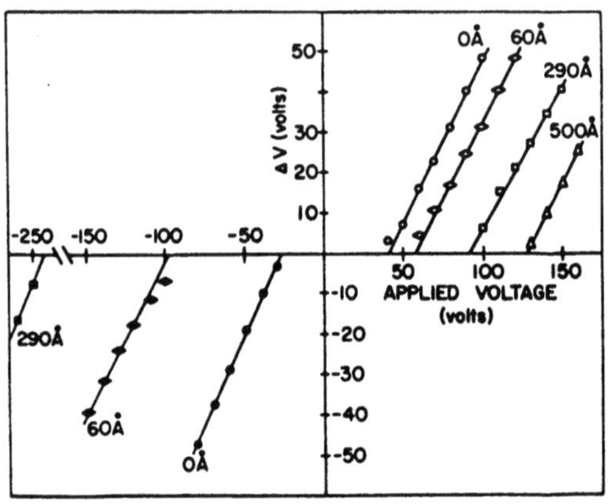

Figure IV-11. Voltage shift ΔV (due to room-temperature trapping) vs applied voltage for metal-nitride-oxide-silicon capacitors with various thicknesses of thermal oxide under 0.25 μ of silicon nitride [(111) oriented 1.4×10^{16} cm^{-3} p-type silicon.] (17)

The main features of the trapping observed in metal-nitride-oxide-silicon structures under positive bias have been explained by Szedon and Chu (18). They propose that the effect is due to electrons tunneling from the silicon conduction band into the oxide conduction band where they drift through the oxide into the nitride and are trapped as is illustrated in Figure IV-12.

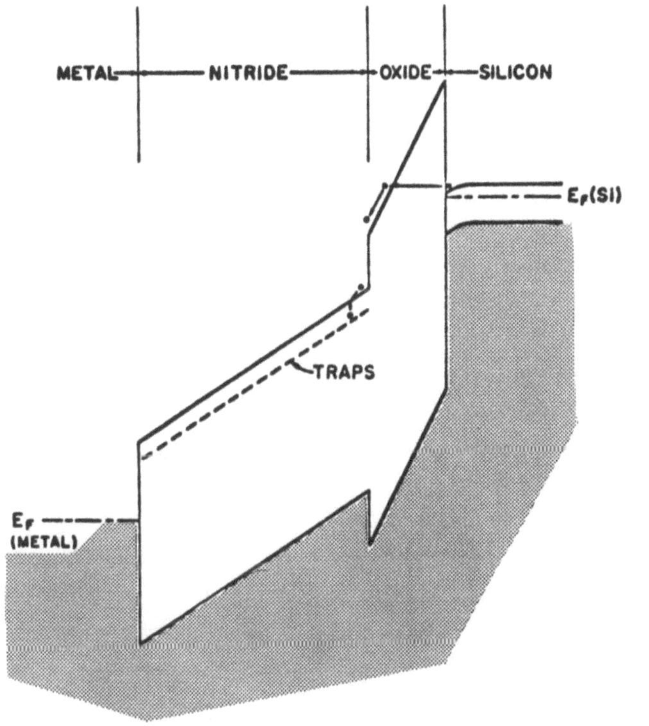

Figure IV-12. Schematic illustration of tunneling model proposed by Szedon and Chu (18) for trapping in metal-nitride-oxide-silicon structures.

The threshold voltage corresponds to the establishment of a field in the oxide large enough to allow tunneling into the conduction band. When the applied voltage is raised above the threshold, enough charge will flow through the oxide and be trapped to reduce the oxide field back to the threshold value. The larger threshold voltages which have been observed for MNOS structures under negative gate bias are presumably due to a lower tunneling probability for holes into the valence band of SiO_2 than for electrons into the conduction band.

Swaroop and Schaffer (19) prepared composite layers of vapor deposited (SiH_4/NH_3) nitride (≃ 4000 Å) and oxide (≃ 1200 Å) and measured their room temperature current-voltage characteristics. They observed that conduction is bulk-limited and conforms to the Frankel-Poole electronic transport mechanism and is due to nitride layers only. The I-V curves of the nitride vs nitride-oxide composites are compared in Figure IV-13.

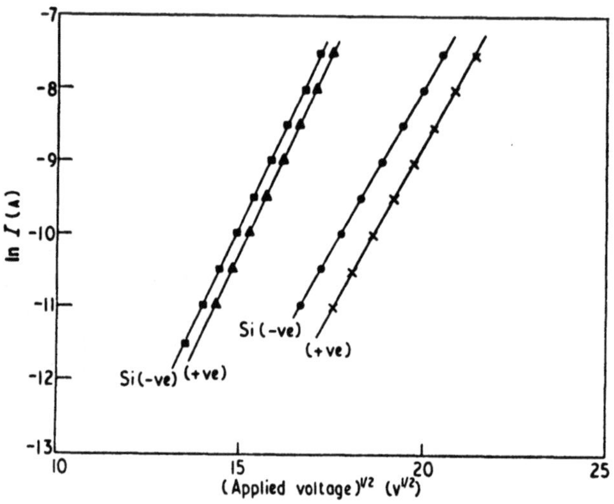

Figure IV-13. ▲ ■ nitride (≃ 2500 Å). x ● nitride (≃ 4000 Å) + oxide (≃ 1200 Å). (19)

The silicon dioxide was grown on Si substrates by the direct oxidation of the Si in a wet oxygen ambient at 1000°C. Before depositing the nitride over the oxide, the latter was annealed at 1000°C in a dry N_2 atmosphere for one hour. Aluminum dot electrodes were evaporated over the nitride and onto the Si back electrode to make uniform electrical contact. These devices were annealed then under a flowing N_2 atmosphere for 5 minutes at 500°C. Table IV-4 summarizes the capacitance and dielectric constant of the structures evaluated.

Table IV-4. Capacitance and Dielectric Constant of MIS Structure at 1 MHz (19)

Composite structure	Capacitance (pF)	Dielectric constant E
MNS ($t \simeq 2500$ Å)	96–102	6·0–6·3
MNOS ($t \simeq 4000$ Å + 1200 Å)	39–43	5·0–5·4
MOS ($t \simeq 2500$ Å)	62–65	3·8–4·0

t is the total thickness of the insulating layer between the metal and semiconductor

SILICON NITRIDE-SILICON DIOXIDE COMBINATIONS
REFERENCES

1. CHU, T.L. et al. Films of Silicon Nitride-Silicon Dioxide Mixtures. ELECTROCHEM. SOC., J., v. 115, no. 3, Mar. 1968. p. 318-322.

2. BROWN, D.M. et al. Properties of $Si_xO_yN_z$ Films on Si. ELECTROCHEM. SOC., J., v. 115, no. 3, Mar. 1968. p. 311-317.

3. DOO, V.Y. and D.R. KERR. Investigation of Refractory Dielectrics for Integrated Circuits. INTERNATIONAL BUSINESS MACHINES CORP., Report No. NASA-CR-995. Mar. 1968. 73 p.

4. RAND, M.J. and J.F. ROBERTS. Preparation and Properties of Some Silicon Oxynitride Films. Paper presented at 1968 Meeting Electrochemical Society, Boston. ELECTROCHEM. SOC., J., v. 115, 1968. p. 194C.

5. FRANK, R.I. and W.L. MOBERG. Reactively Sputtered Silicon Oxynitride as a Dielectric Material for Metal-Insulator-Metal Capacitors. ELECTROCHEM. SOC., J., v. 117, no. 4, Apr. 1970. p. 524-529.

6. KENDALL, E.J.M. The Stabilization of Silicon Surfaces Using Silicon Nitride. BRITISH J. OF APPLIED PHYS., (J. OF PHYS., D), Ser. 2, v. 1, 1968. p. 1409-1420.

7. FORGENG, W.D. and B.F. DECKER. Nitrides of Silicon. AIME METALLURGICAL SOC., TRANS., June 1958. p. 343-348.

8. DRUM, C.M. and M.J. RAND. Chemical Vapor Deposition of a Low-Stress Insulating Film on Silicon. Paper presented at 1968 Electrochemical Society Meeting. ELECTROCHEM. SOC., J., v. 115, 1968. p. 194C.

9. TOMBS, N.C. et al. Solid Solution in the Silicon Nitride-Silicon Dioxide System. ELECTROCHEM. SOC., J., v. 116, no. 6, June 1969. p. 862-863.

10. JONES, R.E. and V.Y. DOO. Integrated Circuit Isolation with Silicon Nitride. ELECTROCHEM. TECHNOL., May-June 1967. p. 297-298.

11. SAVAGE, J.A. et al. Investigation of Silicon Nitride Thin Films for Use in MOST Devices. Royal Radar Establishment (Malvern, Eng.). In its RRE NEWSLETTER AND RESEARCH REVIEW, v. 7, 1968. p. 15/1-2. NASA N69-28679.

12. CHU, T.L. et al. Preparation and C-V Characteristics of $Si-Si_3N_4$ and $Si-SiO_2-Si_3N_4$ Structures. SOLID STATE ELECTRONICS, v. 10, Sept. 1967. p. 897-905.

13. FROHMAN-BENTCHKOWSKY, D. and M. LENZLINGER. Charge Transport and Storage in Metal-Nitride-Oxide-Silicon (MNOS) Structures. J. OF APPLIED PHYS., v. 40, no. 8, July 1969. p. 3307-3319.

14. WANG, P. and N. VanBUREN. Dielectric Films as Crosspoints in Switching Matrices. SYLVANIA ELECTRIC PRODUCTS, Report No. RADC-TR-69-185, July 1969. 73 p. AD 692 170.

15. SWAROOP, B. D-C Conduction in Silicon Nitride and Silicon Nitride-Oxide Composites. Abstract No. 7. The Electrochemical Society, Extended Abstracts of Spring Meeting, Los Angeles, May 10-15, 1970. p. 26-28.

16. SCHAFFER, P.S. and B. SWAROOP. Vapor Phase Growth and dc Breakdown of Silicon Nitride Films. CERAMIC BULLETIN, v. 49, no. 5, 1970. p. 536-538.

17. DEAL, B.E. et al. Electrical Properties of Vapor-Deposited Silicon Nitride and Silicon Oxide Films on Silicon. ELECTROCHEM. SOC., J., v. 115, no. 3, Mar. 1968. p. 300-307.

18. SZEDON, J.R. and T.L. CHU. Paper presented at IEEE Solid State Device Research Conference, Santa Barbara, Calif., June 19-21, 1967.

19. SWAROOP, B. and P.S. SCHAFFER. Conduction in Silicon Nitride and Silicon Nitride-Oxide Films. J. OF PHYS., D, v. 3, no. 5, May 1970. p. 803-806.

MIX
Papier aus verantwortungsvollen Quellen
Paper from responsible sources
FSC® C105338

If you have any concerns about our products,
you can contact us on
ProductSafety@springernature.com

In case Publisher is established outside the EU,
the EU authorized representative is:
Springer Nature Customer Service Center GmbH
Europaplatz 3, 69115 Heidelberg, Germany

Printed by Libri Plureos GmbH
in Hamburg, Germany